AUDREY COHEN COLLEGE LIBRARY
75 Varick St. 12th Floor
New York, NY 10013

Jobs & Economic Development

Robert P. Giloth
Editor

HD
5708.85
U6
J63
1998

Jobs & Economic Development
Strategies and Practice

SAGE Publications
International Educational and Professional Publisher
Thousand Oaks   London   New Delhi

Copyright © 1998 by Sage Publications, Inc.

All rights reserved. No part of this book may be reproduced or utilized in any form or by any means, electronic or mechanical, including photocopying, recording, or by any information storage and retrieval system, without permission in writing from the publisher.

For information:

SAGE Publications, Inc.
2455 Teller Road
Thousand Oaks, California 91320
E-mail@sagepub.com

SAGE Publications Ltd.
6 Bonhill Street
London EC2A 4PU
United Kingdom

SAGE Publications India Pvt. Ltd.
M-32 Market
Greater Kailash I
New Delhi 110048 India

Printed in the United States of America

Library of Congress Cataloging-in-Publication Data

Main entry under title:

    Jobs and economic development: Strategies and practice / edited by
      Robert P. Giloth.
       p. cm.
      Includes bibliographical references and index.
      ISBN 0-7619-0913-3 (alk. paper). — ISBN 0-7619-0914-1 (pbk.:
    acid-free paper)
      1. Hard-core unemployed—United States—Congresses. 2. Urban
    poor—Employment—Government policy—United States—Congresses.
    3. Job creation—Government policy—United States—Congresses.
    4. Occupational training—Government policy—United States—
    Congresses. 5. Community development, Urban—United States—
    Congresses. 6. Economic development projects—United States—
    Congresses. 7. Local government—United States—Congresses. I.
    Giloth, Robert.
    HD5708.85.U6J63 1998
    331.12′042′0973—dc21   97-45332

98 99 00 01 02 03 04 8 7 6 5 4 3 2 1

Acquiring Editor:	Catherine Rossbach
Editorial Assistant:	Renee Piernot
Production Editor:	Sanford Robinson
Production Assistant:	Lynn Miyata
Designer/Typesetter:	Danielle Dillahunt
Cover Designer:	Candice Harman
Indexer:	Virgil Diodato
Print Buyer:	Anna Chin

To Rob Mier—friend and colleague

Contents

Foreword: Jobs—The Real Goal   xi
Norman B. Rice

Acknowledgments   xv

1. Jobs and Economic Development   1
Robert P. Giloth

PART I. BACKGROUND

2. Labor Market Restructuring and Workforce Development: The Changing Dynamics of Earnings, Job Security, and Training Opportunities in the United States   19
Bennett Harrison and Marcus Weiss

3. Why Local Economic Development and Employment Training Fail for Low-Income Communities   42
Elizabeth J. Mueller and Alex Schwartz

4. Networks, Sectors, and Workforce Learning   64
Laura Dresser and Joel Rogers

PART II. TECHNIQUES

5. **Regional Economic Analysis to Support Job Development Strategies**   85
   Brian Bosworth

6. **Labor Market Profiling: Case Studies of Information-Gathering Techniques for Employment Projects**   105
   Peggy Clark and Amy J. Kays

PART III. EXAMPLES

7. **Start-Ups and Replication**   123
   Steven L. Dawson

8. **Ready for Work: The Story of STRIVE**   137
   Lorenzo D. Harrison

9. **New Careers Revisited: Paraprofessional Job Creation for Low-Income Communities**   152
   Janice M. Nittoli and Robert P. Giloth

PART IV. ISSUES

10. **Community Analysis and Organizing for Jobs**   179
    Thomas R. Dewar

11. **The Politics of Jobs in Maine**   195
    Pierre Clavel and Karen Westmont

12. **Changing the Constraints: A Successful Employment and Training Strategy**   214
    Brenda A. Lautsch and Paul Osterman

13. **Prospects for Job-Centered Economic Development**   234
    Robert P. Giloth

Index   249

About the Authors   259

Foreword

Jobs—The Real Goal

Jobs and economic development are inextricably linked. One is impossible without the other. Economic development creates jobs, and jobs provide the disposable income that further fuels the economy.

The U.S. economy has finally emerged from a prolonged malaise. The Pacific Northwest is outperforming many regions in the nation, with Boeing, Microsoft, Starbucks, and Nordstrom all thriving and claiming larger shares of their respective national and international markets. Seattle's proximity to the Pacific Rim has made international trade an even more important part of the regional economy; today, one out of every four jobs in Washington state depends on international trade. This prosperity is helping to make Seattle, the "Emerald City," even greener.

Some may say that now is hardly the time to worry about economic development. But the pendulum will swing back in the opposite direction, sooner than any of us would want it to. Now is the time to seize the opportunity to do what we have long held as a lofty goal—to spread prosperity throughout the entire community and to help people lift themselves out of the mire of poverty.

Economic development efforts at present have been commendable, and we can be rightly proud of many successes. Unfortunately, however, economic development has too often been distracted from what should be its singular goal. Too often, the focus has been on the means and not the end. Success is declared when the grant has been won, or the loan made, or the savior company announces that it will locate in a particular community. These events are important, but they are only the beginning of a process that must ultimately end in the creation of jobs that pay enough to support a family and help low-income people significantly raise their standard of living.

Yes, job creation is nearly always the by-product of economic development. But today it is not enough to be just a by-product. In my mind, the creation of jobs that pay a proverbial family wage must be the strong, clear goal of our efforts.

Some may argue that fostering entrepreneurship or diversifying the local economic base will create prosperity across the broadest cross section of society. I hope we've learned a lesson, though, from the failed "trickle-down" theory of economics, which left us feeling "trickled on." Left to its own devices, the market has never created broad-based prosperity, or even opportunity. Instead, it has tended to widen further the gap between the haves and the have-nots. As a society, we must ask ourselves whether we can be satisfied with a system that consigns ever more children to lives of poverty. How can we be satisfied with notions of economic progress that are not judged by the health of our community at large?

As mayor of Seattle from 1990 to 1998, I have strived to make social equity and economic opportunity hallmarks of my administration. In Seattle, we did our economic development work well. We forged a remarkable public-private partnership that helped make our downtown one of the healthiest and most vibrant in the nation. We increased access to capital for small businesses and made tax and regulatory policies more business-friendly. Infrastructure investments bolstered our port and industrial areas. We even focused new public investments in our poor neighborhoods. Despite these efforts and a robust economy, we still saw pockets of poverty and unemployment in some communities three times higher than average.

Today, we are tackling this most difficult and stubborn challenge through an innovative partnership known as the Seattle Jobs Initiative. With support from local and national foundations and a major investment of city funds, we are fundamentally changing workforce development. We are working with employers in our growth sectors to link low-income residents with jobs that pay a livable wage, jobs that are found not only in poor neighborhoods, but within the overall regional economy. We focus on wage levels and job retention, and on being responsive to the needs of both the job seeker and the employer. Our holistic approach includes creation of the Ready to Earn fund, which can be tapped by both employers and new workers to pay for such things as training, transportation, child care, even emergency housing—all obstacles that can stymie new workers.

Welfare reform was not the sole motivating factor behind the Seattle Jobs Initiative, but this program will help ease the impact of that reform on the most vulnerable of our community, and ease the burden on our food banks, homeless shelters, and public health clinics. The significance of this effort cannot be overstated.

I have high hopes for the Seattle Jobs Initiative and similar efforts I see emerging in other cities across the nation. I am proud to see the focus in Seattle and elsewhere finally being placed on real, long-term outcomes of economic development. I know that success will not be easy, and real change will not come overnight. But I believe that success can be achieved with dedication and hard work. If our community can improve the economic conditions of all people, then I believe we will have created a vision and a plan for a just and civil society for the 21st century.

—NORMAN B. RICE
Former Mayor,
City of Seattle

Acknowledgments

This book grew out of my work designing and now implementing a multicity jobs demonstration. Although the contents of the book do not necessarily reflect the opinions of the Annie E. Casey Foundation, I want to thank Susan Gewirtz, James Hyman, Sandy Jibrell, Doug Nelson, and Ralph Smith for their ongoing support and interest. I also want to thank a list of people too long to recount here, community organizers, jobs practitioners, researchers, foundation colleagues, and policy makers who have shaped my ideas about reforming workforce projects and systems. Above all, chapter authors have produced important and timely contributions. Thanks to Cassandra Maxey for helping with author communications and to Cinder Hypki for helping me shape the final manuscript. Finally, my appreciation and love to Kari, Emma, and Lian for putting up with the inevitable aggravation caused by deadlines and weekend work.

1

Jobs and Economic Development

ROBERT P. GILOTH

In March 1997, 100 people convened at the Watergate Hotel in Washington, D.C., to grapple with the issues of job availability, livable wages, and welfare changes (Jobs for the Future 1997). Sponsored by Jobs for the Future, a nonprofit consulting organization based in Boston, the conference examined the research techniques, findings, and policy implications of the Jobs Gap study undertaken by the Minnesota JOBS NOW Coalition (Steuernagel 1995).[1] This provocative study concluded that there were 31 people in Minnesota seeking every livable wage job requiring a year of training or less.[2] In other words, not enough good jobs existed at a time when public policy was pushing more and more people, many with minimal preparation, into the labor market. Studies completed in Illinois and Wisconsin identified similar jobs gaps (Lafer 1994; Carlson and Theodore 1995). The JOBS NOW Coalition launched an advocacy agenda based upon the Jobs Gap study results, attacking state-level corporate welfare policies, promoting livable wage ordinances, and advocating improved labor market information.

The March 1997 conference explored the implications of the jobs gap approach for an understanding of today's employment problems and potential solutions. On the opening panel, Frank Doyle, retired executive vice president of General Electric, emphasized the critical role of education reform in remedying the income inequality now characterizing the U.S. economy. While acknowledging runaway CEO salaries, he argued that such salaries are the wrong reform target. Karen Nussbaum of the AFL-CIO and founder of 9 to 5, a labor organization for female clerical workers, argued, in contrast, for the countervailing social force of a revived labor movement to bargain for more equitable arrangements between workers and the owners of wealth. She observed that today's clerical workers, despite their making the same wages such workers made 20 years ago, or even lower wages, now need to know several computer languages.

As the conference unfolded, other dimensions of the jobs and economic development challenge emerged. The impacts of the Personal Responsibility and Work Opportunity Reconciliation Act of 1996 on welfare recipients and children gave some urgency to the discussion.[3] Human service practitioners underscored the problems of illiteracy, alcohol and drug dependence, and lack of job readiness that prevent many inner-city job seekers from obtaining and keeping jobs. Business practitioners gave examples of unfilled livable wage jobs, global recruitment of machinists, and unfilled training/apprenticeship programs. And there was recognition among conference participants that many employment training approaches do not work.

The multiple viewpoints and constituencies represented at the Jobs Gap conference illustrate the intellectual and practical terrain involved in improving labor market prospects for low-income communities. Despite overall U.S. job growth, new labor market policies and approaches are needed at the local, regional, and national levels because of changing labor markets, the growing income inequality and size of the working poor population, the continued exclusion of low-income and minority job seekers from good jobs, and the limitations of employment training and economic development strategies in alleviating poverty.

This book explores and documents local "laboratories" of jobs innovation that have improved connections between poor communities and livable wage jobs, creating, in a sense, new labor market intermediaries. I call this set of practices *job-centered economic development,* because of its disciplined focus on good jobs as an outcome for low-income communities and because it takes advantage of and has benefits for demand-side economic development strategies (Giloth 1995).[4] This eclectic group of strategies and projects combines education, human services, economic development, and employment training. It focuses on identifying and accessing good jobs, networking among employers, building career ladders, enabling job retention, and advocating policies in support of livable wage jobs (Fitzgerald 1993; Giloth 1995). Although somewhat awkward, the term *job-centered economic development* underscores jobs outcomes, not simply economic adaptation, new business generation, job readiness, or workforce attachment.[5]

An array of projects located in different regional, institutional, and market contexts represent job-centered economic development, including short- as well as long-term training, enterprise development, and sectoral initiatives. The emergent quality of many of these innovations, as well as their focuses on particular places, makes their description and analysis difficult. Consequently, in order to advance our understanding of these innovations, this book brings together academics and practitioners to reflect upon the contexts and practices of job-centered economic development from various design, practice, evaluation, and policy perspectives.

From a broader perspective, job-centered economic development suggests the formulation of a new "social contract" among business, government, labor, and community around access to good jobs and the need for more integrated labor market policies and institutions at all levels of government and the economy.[6] Unfortunately, employment policy is fragmented in the United States, with job quality de-emphasized and the authority to design and implement jobs-related programs increasingly devolving to the states.[7] Although there have been calls for more aggressive federal policies to ensure good jobs, contemporary budget politics and pressures to return more power to the states make such redistributive programs improbable (Blank 1997). What this suggests for the practice of job-centered economic development is a national policy environment characterized by both barriers and opportunities—one that in the short run, however, limits macro interventions to influence the overall supply of good jobs as identified at the Jobs Gap conference.

In the rest of this introductory chapter, I will summarize the assumptions and practices of job-centered economic development.[8] I begin with an overview of labor market outcomes and barriers for low-income communities, and then focus on describing in more detail the practice of job-centered economic development, illustrative jobs projects, and the challenges facing this field. I conclude the chapter with an interpretive guide to the parts and chapters of this book.

Labor Market Outcomes and Barriers

Despite a national unemployment rate of less than 5% in 1997, and more than 12 million jobs created since 1993, labor market outcomes have stagnated or worsened for many U.S. workers, particularly in communities of color. Our job machine is working, but it is not creating enough good jobs that are accessible to members of low-income communities. Moreover, the "lean" and flexible production of the new economy—embodied, for example, in increased contingent work—is creating additional employment barriers for these populations (Harrison 1994; Cappelli et al. 1997).

Decreasing unemployment has not resulted in commensurate reductions in poverty. Family poverty rates of 14% or more have existed for the past 15 years; the rate is 30% for African Americans. Black and Hispanic unemployment rates doubled that of Whites in 1995, rising to 10.4 and 9.3, respectively. Out-of-school youth between the ages of 16 and 24 without a high school diploma suffer close to 50% unemployment (U.S. Department of Labor 1994). In 1995, 13.5 million workers, or 10.1% of the civilian labor force of 132 million, could be classified as underemployed, including the unemployed, discouraged or marginally connected workers, and involuntary part-time workers (Mishel, Bernstein, and Schmitt 1997).

The proportion of workers in poverty grew some 20% between the late 1970s and 1995, numbering close to 4 million families with full-time or part-time workers (Lazere 1997). Often without health benefits and buffeted by the instability of contingent work, the working poor represent the likely outcome of the Personal Responsibility and Work Opportunity Reconciliation Act of 1996. In 1995, 5.8 million children lived in families in which one or more parents worked at least 50 weeks of the year but did not make above-poverty-level incomes. This represented an increase from 3.4 million children in working poor families two decades ago. Between 1989 and 1995, the number of children in such families increased by 35% (Annie E. Casey Foundation 1996).

Declining wages, incomes, and returns to education characterize worsening labor market outcomes for many Americans, although there are differences by age, race, and gender. In 1995, real median family income was $1,438 less than in 1989. In 1995, the real wages of male high school graduates were 27% below those of 1979; women's wages were 19% below 1979 levels (Mishel, Bernstein, and Schmitt 1997). A Black/White wage differential exists at all levels of education: Black high school dropouts earn 86% of the wages earned by White dropouts, and Black high school grads earn only 82% of what White high school graduates earn. The situation is worse for Black males (Jeffries and Schaffer 1996). A recent California study found that, over time, better-educated minority workers earned less than Whites of comparable or lesser educational achievement (Carnoy and Rothstein 1996).

The current economy and labor market produce too few good jobs, lower wages, and heightened economic insecurity. Demand-side explanations emphasize the effects of global competition on job structuring and downward wage pressures, increases in outsourcing and temporary work, and structural shifts in the economy from manufacturing to services. Technological changes complicate the impacts of these processes (Blank 1997). Demand-side forces have shaped demand for higher skills as well as lowered the skills required in many jobs; they have also influenced the trends toward preoccupation with soft skills, deunionization, lower wages, and the regional and global reordering of economic activity.

In the past, dual labor market theory examined the connections between good and bad jobs and the relegation of the poor to bad jobs. Regional labor markets of today are highly segmented, and flexible production methods and organization have collapsed—or made more complex—distinctions between primary and secondary sectors, or core and periphery firms (Peck 1996). Prominent barriers preventing access even to lower-paying jobs as well as undermining career development are related to skills, geography, informal hiring networks, discrimination, career ladders, and broken or fragmented public systems.[9]

Skills

Investment in the long-term acquisition of skills is paramount for U.S. competitiveness as well as for the future of low-income communities (U.S. Department of Labor 1991; Ferguson 1995). Illiteracy, innumeracy, lack of critical thinking skills, and deficiency in computer know-how are reflected in lower wages and earnings (U.S. Department of Labor 1994). Only 5-10% of job openings in central cities can be filled by workers with few cognitive skills and spotty work records (Holzer 1996). Some economists, however, question the singular focus on skills, noting some firms' use of skill inflation to exclude certain workers, lower returns on education, and the decline in wages and incomes regardless of skills (Harrison and Weiss 1998). Moreover, a seeming anomaly is that many manufacturers prefer to hire lower-skilled Hispanics over higher-skilled Blacks (Waldinger forthcoming).

There has been a dramatic increase in the demand for what are known as *soft skills,* a collection of behavioral skills related to motivation, teamwork, and problem solving. The unfortunate shorthand for these skills is *work ethic.* This term is unfortunate not only because of its labeling function, but also because it fails to recognize such demand-side factors as the growth of retail, team production processes, and emphasis on quality service. The soft skills mismatch is most profound for inner-city Black males, who are frequently labeled by employers, including minority employers, as angry, disruptive, and unwilling to adopt business norms (Wilson 1996; Moss and Tilly 1996b).

Geography

The "spatial mismatch" hypothesis posits that geographic barriers separate job-seeking populations from where the jobs are: Job seekers are in inner-city communities, and newly created jobs are in the suburbs (Kain 1992). Geographic barriers exist because of investment choices, racial segregation, low inner-city auto ownership, and the inadequacy of public transportation systems to reach new centers of job growth (Hughes 1993). At the same time, many central-city jobs are in retail and service and have higher hard and soft skill requirements compared with low-skilled manufacturing and warehouse jobs in the suburbs (Holzer 1996). And inner-city residents frequently fail to gain access to jobs in the vicinity of their residences (Kasinitz and Rosenberg 1993; Immergluck 1997; Newman and Lennon 1995). Consequently, the spatial mismatch hypothesis may be too narrow; the real issue is the social and spatial isolation of low-income communities from access to economic opportunity (Goldsmith and Blakely 1992).

Informal Hiring Networks

Existing employees and their networks of family, friends, and associates serve as primary recruitment mechanisms for many firms (Granovetter 1995; Waldinger forthcoming). Firms turn to existing employees for referrals to fill job openings, and employees turn to people they know or with whom they are acquainted. Several factors shape this recruitment strategy. As firms become disconnected from neighborhoods and cities, they lose the ability to source their jobs formally as well as to discern who will be productive workers. This also reflects their lack of confidence in public training and job matching systems such as those offered by state employment services. Given that job search and job turnover have economic costs, and that firms may have their own discriminatory or soft skills lenses, it is easier and cheaper for employers to ask "trusted" and "successful" employees to make referrals. Employees know what jobs require and have a self-interest in recommending successful candidates. In some cases, firms award bonuses (or bounties) to employees who help fill empty positions through referrals.

What this means for many inner-city job seekers is disconnection from jobs in particular firms because they are not in the networks of the firms' employees (Holzer 1996). There is some evidence that "formality" in hiring methods increases the probability that Blacks will be hired (Moss and Tilly 1996a). Networks can enable substantial local hiring once subgroups gain access; this has occurred for the Hispanic and Asian populations in some regions because of their easier access into specific industries (Waldinger 1996, forthcoming). However, the increased use of temporary service agencies may reflect a lack of confidence by firms in the ability of even informal hiring networks to source successful job candidates.

Career Ladders

Simply put, in order to achieve lean and flexible production, firms have dismantled the career ladders that used to represent fairly transparent pathways to more responsible and higher-paying positions (Harrison 1994). Structural shifts in the economy away from manufacturing have also contributed to this redefinition of internal labor markets, in part through the dismantling of apprenticeship programs and training partnerships with community colleges. Diminished union representation has exacerbated this loss. Although career ladders still exist, they are less transparent, operating across firms, sectors, and regions.

Discrimination

Discrimination against minority job seekers prevails in a range of occupations and jobs. Although many gains have been achieved in opening up jobs, Blacks remain overrepresented in menial occupations and underrepresented in personal service occupations, such as waiter or dental hygienist (Hacker 1992, 111). Despite low skill requirements, Blacks made meager inroads into the construction trades, and their higher-paying jobs, even though the industry in the case of New York City grew 63% between 1980 and 1990 and there were many highly publicized efforts to open up the trades (Waldinger 1996). Recent discrimination testing for jobs confirms that African Americans remain at a disadvantage in the labor market even when they have comparable skills and experience; in employment audit tests, they experienced discrimination in up to 20% of matched interviews (Bendick, Jackson, and Reinoso 1994). Discrimination against Blacks also varies between city and suburbs, with suburban firms less likely to hire Blacks even though they may be a better fit with jobs. Some firms also prefer hiring Hispanics or Black females for jobs rather than Black males (Holzer 1996).

Fragmented/Ineffective Training Systems

A 1994 survey of local economic development officials by the National League of Cities reported that few recognized poverty alleviation as part of their mission and practice (Furdell 1993). In general, economic developers are loath to mix private sector job creation with social services and poor people; they fear sending the wrong signals about the local business climate (Stillman 1994).

The employment and training system is fragmented, uneven in its effectiveness, and underinvested in innovation. In many metropolitan regions, multiple service delivery areas and Private Industry Councils rarely coordinate their investments of federal employment and training funds. National evaluations of employment training programs under the Job Training Partnership Act have reported a rather dismal record of employment training results, especially for men and young people (Orr et al. 1995). An obvious gap in job training investment lies in the area of postplacement support, those mechanisms that increase job retention, which is the real outcome of employment training (Whiting and Taylor 1997). Although exceptions exist, urban community colleges, repositories of considerable training money and expertise, have been ineffective in integrating basic skills, customized training, community-based literacy efforts, and employers for low-income communities (Fitzgerald and Jenkins 1997). Finally, private employers in the United States have underinvested in the training of their workers, particularly in entry-level positions, compared with employers in other advanced economies (Berg 1996).

Local Labor
Market Problem Solving

At the same time labor market outcomes were stagnating or worsening for many inner-city communities, examples of job-centered economic development emerged around the United States. These projects made job connections between low-income people and family-supporting jobs, engaged employers and communities in novel ways, and employed educational practices that were contextual, self-paced, and accessible. Their results were better than those of most employment and training programs and showed poverty alleviation effects (Giloth 1995; Dewar and Scheie 1995; Clark and Dawson 1995: Harrison, Weiss, and Gant 1995; O'Regan and Conway 1993).

The Center for Employment Training (CET), based in San Jose, California, conducts arguably the best-known 6- to 8-month skills training program in the past decade for such occupations as medical assistant, machine operator, and printer (Melendez 1996; Harrison and Weiss 1998). Starting as an opportunity industrial center in the early 1970s in the Silicon Valley, CET developed unique ties to employers and the community while embracing a host of classroom innovations and committing itself to lifetime service to its graduates. CET reinvented itself several times in the 1970s and 1980s as federal funding changes swept through the employment training field, becoming certified by the state as a vocational tech school, for example, so that it could take advantage of Pell Grants. By the 1990s, CET was known for providing one of the few adult training programs that showed not only positive economic returns to participants but poverty reduction as well (Orr et al. 1995). As Russ Tershy, longtime president of CET, often reminds people, "Graduation from CET means getting a job."

Recent case studies of CET have emphasized how it became "embedded" in community and employer networks (Melendez 1996; Harrison and Weiss 1998). This embeddedness enabled CET to function as an effective labor force intermediary. On the one hand, CET grew out of faith-based participation in the farmworkers' movement in California; it was part of the struggle of Mexican Americans for civil rights and just wages. On the other hand, CET drew upon the boom and bust of Silicon Valley to develop relationships with employers who helped design occupations and curricula, provide teachers and equipment, and ultimately provide jobs. During the 1980s, CET established training sites in other parts of California.

In 1993, CET began replicating in East Coast and Midwest cities with funding from the U.S. Department of Labor after three rigorous evaluation studies documented its impacts. Since 1996, the Local Initiatives Support Corporation (LISC), a national housing intermediary, has joined the effort to replicate CET in cities in which LISC had major community development investments. CET replicates in one of two ways: by setting up satellites under its management or by franchising to local employment trainers and providing technical assistance. Although evaluations of the replications are not yet completed, early results raise cautionary questions about how to replicate "embeddedness," how to find the right social entrepreneurs, and how to navigate the constraints and preferences of local policy environments that compromise the CET model (Melendez 1996).

Despite unanswered questions about replication, CET has clearly set the standard for effective workforce intermediaries. It is market oriented in its engagement with employers and in the design of training; it integrates human services, economic development, and training; and it is highly networked or embedded in communities and local economies, "spanning the boundaries" that separate labor market actors (Harrison and Weiss 1998). Moreover, CET combines discipline and an entrepreneurial focus on results (Melendez 1996).

Many other examples of job-centered economic development share these CET attributes (Giloth 1995). In contrast to CET, however, these jobs projects focus on different target groups of firms and job seekers; use other forms of intervention, such as enterprises, "tough love" job readiness, or supported work; and overcome barriers such as transportation. Their origins are also quite diverse, and include community development, civil rights, and human services. Examples are Focus Hope in Detroit, Michigan; Cooperative Home Care Associates in the South Bronx, New York; Suburban Job Link and the Chicago Manufacturing Institute; Project QUEST in San Antonio, Texas; the Asian Neighborhood Design Center in Oakland, California; and Pioneer Human Services in Seattle, Washington (Giloth 1995; Dewar and Scheie 1995; Clark and Dawson 1995; Harrison, Weiss, and Gant 1995; O'Regan and Conway 1993).

Examining these jobs projects underscores four additional characteristics of job-centered economic development. First, effective jobs projects sharply target their interventions to specific groups of job seekers, occupations, and employers (Clark and Kays 1996). One size does not fit all. Job seekers differ by race, gender, ethnicity, job readiness, literacy, aptitude, aspiration, self-sufficiency, and geographic location. Second, many jobs projects develop deep knowledge about sectors, niches in the economy with shared products and processes, and become valued players among firms within the targeted sectors (Clark and Dawson 1995). Third, jobs projects are regional; they extend beyond neighborhood and relate to the regional economy while not leaving behind their important ties to place (Dewar and Scheie 1995). Finally, job-centered economic development reforms how labor markets function for low-income communities (Clark and Dawson 1995; Okigaki 1997).

Although the jobs projects of this generation demonstrate the focus and integration needed for effective labor market intermediaries, they also have limitations. They are frequently small, fragmented, and disconnected from public systems, existing in parallel to them in some cases (Giloth 1995). The challenge for job-centered economic development is to combine effective practice with replication, scaling up, and systems change.

The Aspirations and Organization of This Book

This book brings together theoreticians, analysts, and practitioners to share their work on job-centered economic development, exploring its assumptions, design tools, case studies, policies and politics, and evaluation challenges. The book represents, in a sense, a conversation about a set of emergent practices, a dialogue necessarily incomplete and at times contentious. The challenges of this volume are to make these various voices and perspectives cohere to create a richer understanding of "good jobs" projects and to provide direction for those who must confront the substantial labor market barriers that face inner-city communities.

The chapters in Part I examine contemporary labor markets, evaluation research about what works for low-income communities, and promising theories of sectoral employment planning. In Chapter 2, Harrison and Weiss examine the state of labor markets, with particular attention to earnings deterioration, the skills debate, and the implications of networking for the design of employment interventions. In Chapter 3, Mueller and Schwartz critically review the effectiveness of conventional economic development, welfare, and employment training approaches to improving jobs prospects for low-income communities, arguing that successful employment interventions must be comprehensive. In Chapter 4, Dresser and Rogers articulate an ambitious version of sector-based employment and economic development planning with examples drawn from their work in Wisconsin.

Part II contains two how-to chapters on critical planning dimensions of job-centered economic development. In Chapter 5, Bosworth discusses how to analyze regional economies in terms of existing jobs, sectors, and clusters, and how to get beyond secondary data sources to talk with companies. Clark and Kays, in Chapter 6, present examples of how to target jobs projects to groups of unemployed/underemployed residents during planning and implementation.

Part III presents reports on three jobs projects that demonstrate different strategies related to soft skills, career ladders, postplacement services, enterprise formation, and replication. Dawson, in Chapter 7, shares the history of Cooperative Home Care Associates, a successful worker-owned company in the South Bronx that has replicated in Boston and Philadelphia and is adapting to the current managed-care environment. In Chapter 8, Harrison tells the story of STRIVE, a "tough love" job readiness program started in East Harlem and now replicating nationally. Nittoli and Giloth, in Chapter 9, revisit the New Careers movement of the 1960s and explore the current practice of community outreach workers as an occupation that draws upon the indigenous skills and experiences of community residents.

The contributions to Part IV ask more pointed questions about the practice, limits, and implications of job-centered economic development. In Chapter 10, Dewar shares how community residents and their organizations can demand and design jobs projects and hold them accountable. Clavel and Westmont, in Chapter 11, examine the politics of jobs in Maine and the tensions among the "good jobs movement," employment hiring agreements, and the business incentive approach to economic development. Lautsch and Osterman, in Chapter 12, present their evaluation study of Project QUEST in San Antonio. Although these researchers did not use a random assignment design, their findings demonstrate large impacts for this distinctive labor market intermediary that provides long-term training.

Finally, I conclude the book in Chapter 13 by extracting themes and questions from the preceding chapters and discussing other implications of recent welfare and workforce legislation. I address the challenge of linking job-centered economic development to broader policy discussions around jobs, cities, and low-income neighborhoods. That is, I reflect on the potential of job-centered economic development to contribute to our understanding of the larger issues related to jobs gaps, income inequality, welfare change, and education reform raised at the outset of this chapter.

Notes

1. Controversy exists among labor market analysts as to the methodology used in this and similar studies. Questions relate to the definition of *livable wage,* the value of cross-sectional estimates of job availability, whether estimates include tax benefits such as the Earned Income Tax Credit, and whether such analyses provide an adequate understanding of the economy for the purposes of policy and advocacy.

2. The terms *livable wage job, family-supporting job,* and *good job* are generally used interchangeably. They refer to jobs that result in above-poverty-level family incomes, have benefits, and allow some career mobility. Nevertheless, there is much variation in how such jobs are defined, with wages ranging from $8.00 to $11.50 an hour, depending upon size of family, government assistance, and tax relief. Discussions about how to define good jobs are not new. Theorists of internal labor markets and subemployment indices and advocates for the Humphrey-Hawkins full employment bill in the 1970s discussed the definition of good jobs in some depth (Mier 1993; Gordon 1979).

3. The Personal Responsibility and Work Opportunity Reconciliation Act of 1996 abolished the 60-year-old system of Aid to Families with Dependent Children, replacing it with block grants to states as well as time restrictions and work requirements. States had to submit plans for Temporary Assistance to Needy Families by July 1, 1997. The work requirements are that 25% of persons on a state's welfare caseload must be working 20 hours a week in FY 1997, and 50% of those on the caseload must be working 30 hours per week by 2002 (see Center for Community Change 1997).

4. Job-centered economic development is not an innovation; antecedents exist in the practice of community development corporations, labor-based development strategies, plant closing fights, union organizing, and even public sector planning such as the Greenhouse Compact in Rhode Island in the early 1980s and the Chicago Works Together development plan of 1984 (Mier, Moe, and Sherr, 1986; Giloth and Moe, forthcoming). It also goes by other names, such as *targeted economic development, employee- or labor-centered economic development,* and *sectoral employment development* (see, e.g., Ranney and Betancur 1992). Job-centered economic development shares common elements with asset-based strategies such as microenterprise and with supported work enterprises that serve homeless and disabled populations (Emerson and Twersky 1996).

5. These projects defy simple categorization as welfare to work, job training, or economic development. They serve multiple groups, combine different strategies, and draw upon a variety of funding streams.

6. Comparisons are frequently made to the labor market planning and adjustment systems of other advanced economies (see, e.g., Kuttner, 1984; Harvey, 1989; Weil, 1992).

7. In fall 1997, Congress was considering once again revamping the employment training system. Final action is not likely until 1998. Whatever legislation passes and is signed will devolve more authority to the states, include block grants, increase local flexibility, and have some version of workforce boards (see, e.g., Baker 1997).

8. This book does not explore several important assumptions about jobs. First, many argue that the issue of jobs is the wrong focus, that our concern should instead be for careers, "employability security," and how to create one's own work (see, e.g., Kanter 1995). The second assumption is that increased income automatically translates into better outcomes for families and children (see, e.g., Mayer 1997; Edin and Lein 1997).

9. A succinct review of these and other labor market barriers, as well as targeted policy prescriptions, is presented by Bluestone, Stevenson, and Tilly (1993).

References

Annie E. Casey Foundation. 1996. *Kids count data book 1996.* Baltimore: Annie E. Casey Foundation.

Baker, S. 1997. *Restructuring the workforce development system: A wide-angle snapshot.* Philadelphia: 21st Century League.

Bendick, M., C. W. Jackson, and V. A. Reinoso. 1994. Measuring employment discrimination through controlled experiments. *Review of Black Political Economy* 23:25-48.

Berg, P. 1996. Training: A plan for all workers. In *Reclaiming prosperity: A blueprint for progressive economic reform,* edited by T. Schafer and J. Faux. Armonk, NY: M. E. Sharpe.

Berry, D. F. Forthcoming. The Jobs and Workforce Initiative: Building new roads to success in northeast Ohio. *Economic Development Quarterly.*

Blank, R. 1997. *It takes a nation: A new agenda for fighting poverty.* Princeton, NJ: Princeton University Press.

Bluestone, B., M. H. Stevenson, and C. Tilly. 1993. Public policy alternatives for dealing with the labor market problems of central city young adults: Implications from current labor market research. Paper prepared for the Social Science Research Council Policy Conference on Persistent Urban Poverty, November 9-10, Washington, DC.

Cappelli, P., L. Bassi, H. Katz, D. Knoke, P. Osterman, and M. Useem. 1997. *Change at work.* New York: Oxford University Press.

Carlson, V. L., and N. Theodore. 1995. *Are there enough jobs? Welfare reform and labor market reality.* Chicago: Illinois Job Gap Project.

Carnoy, M., and R. Rothstein. 1996. *Hard lessons in California: Minority pay gap widens despite more schooling, higher scores.* Working Paper 116. Washington, DC: Economic Policy Institute.

Center for Community Change. 1997. *The new welfare law: A handbook for organizers.* Washington, DC: Center for Community Change.

Clark, P., and S. L., Dawson, with A. J. Kays, F. Molina, and R. Surpin. 1995. *Jobs and the urban poor: Privately initiated sectoral strategies.* Washington, DC: Aspen Institute.

Clark, P., and A. J. Kays. 1996. *Labor market profiling: Information gathering tools and techniques for community-based employment projects.* Washington, DC: Aspen Institute.

Dewar, T., and D. Scheie. 1995. *Promoting job opportunities: Toward a better future for children and youth.* Baltimore: Annie E. Casey Foundation.

Edin, K., and L. Lein. 1997. *Making ends meet: How single mothers survive welfare and low-wage work.* New York: Russell Sage Foundation.

Emerson, J., and F. Twersky. 1996. *New social entrepreneurs: The success, challenge and lessons of non-profit enterprise creation.* San Francisco: Roberts Foundation.

Ferguson, R. F. 1995. Shifting challenges: Fifty years of economic change toward Black-White earnings equality. *Daedalus* 24:37-76.

Fitzgerald, J. 1993. Labor force, education, and work. In *Theories of local economic development: Perspectives from across the disciplines,* edited by R. D. Bingham and R. Mier. Newbury Park, CA: Sage.

Fitzgerald, J., and D. Jenkins. 1997. *Making connections: Best practice in U.S. community colleges in connecting the urban poor to education and employment opportunities.* Chicago: Great Cities Institute, University of Illinois.

Furdell, P. 1993. NLC studies local officials' perception of poverty: Results reflect severity and seriousness of issue. *Nation's City Weekly,* September 13, 5-8.

Giloth, R. P., 1995. Social investment in jobs: Foundation perspectives on targeted economic development during the 1990s. *Economic Development Quarterly* 9:279-289.

Giloth, R. P., and K. Moe. Forthcoming. Jobs, equity, and the mayoral administration of Harold Washington in Chicago (1983-1987). *Journal of Policy Studies.*

Goldsmith, W. W., and J. Blakely. 1992. *Separate societies: Poverty and inequality in U.S. cities.* Philadelphia: Temple University Press.

Gordon, D. M. 1979. *The working poor: Toward a state agenda.* Washington, DC: Council of State Planning Agencies.

Granovetter, M. 1995. *Getting a job: A study of contacts and careers.* Cambridge, MA: Harvard University Press.

Hacker, A. 1992. *Two nations: Black and White, separate, hostile, unequal.* New York: Ballantine.

Harrison, B. 1994. *Lean and mean: The changing landscape of corporate power in the age of flexibility.* New York: Basic Books.

Harrison, B., and M. Weiss. 1998. *Workforce development networks: Community-based organizations and regional alliances.* Thousand Oaks, CA: Sage.

Harrison, B., M. Weiss, and J. Gant. 1995. *Building bridges: CDCs and the world of employment training.* New York: Ford Foundation.

Harvey, P. 1989. *Securing the right to employment: Social welfare policy and the unemployed in the United States.* Princeton, NJ: Princeton University Press.

Holzer, H. J. 1996. *What employers want: Job prospects for less-educated workers.* New York: Russell Sage Foundation.

Hughes, M. A. 1993. *Over the horizon: Jobs in the suburbs of major metropolitan areas.* Philadelphia: Public/Private Ventures.

Immergluck, D. 1997. *Breaking down barriers: Prospects and policies for linking jobs and residents in the Chicago empowerment zone.* Chicago: Woodstock Institute.

Jeffries, J. M., and R. L. Schaffer. 1996. Changes in the labor economy and labor market state of Black Americans. In *The state of Black America 1996,* edited by A. Rowe and J. M. Jeffries. New York: National Urban League.

Jobs for the Future. 1997. *The jobs gap: Where are the living wages?* A national forum convened by Jobs for the Future, Washington, DC, March 27. Boston: Jobs for the Future.

Jobs for the Future, Mt. Auburn Associates, N. Nye, Brandon Roberts and Associates, and R. Schramm. 1996. *Federal jobs policy: History, current status, and future challenges.* McLean, VA: Neighborhood Funders Group.

Kain, J. 1992. The spatial mismatch hypothesis: Three decades later. *Housing Policy Debate* 3:371-462.

Kanter, R. M. 1995. *World class: Thriving locally in the global economy.* New York: Simon & Schuster.

Kasinitz, P., and J. Rosenberg. 1993. Why enterprise zones will not work. *City Journal* 3:63-71.

Kuttner, R. 1984. *Economic illusion: False choices between prosperity and social justice.* Boston: Houghton Mifflin.

Lafer, G. 1994. The politics of job training: Urban poverty and the false promise of JTPA. *Politics and Society* 22:349-388.

Lazere, E. 1997. *The poverty despite work handbook.* Washington, DC: Center on Budget and Policy Priorities.

Mayer, S. E. 1997. *What money can't buy: Family income and children's life chances.* Cambridge, MA: Harvard University Press.

Melendez, E. 1996. *Working on jobs: The Center for Employment Training.* Boston: Mauricio Gastón Institute, University of Massachusetts.

Mier, R. 1993. *Social justice and local development policy.* Newbury Park, CA: Sage.

Mier, R., K. Moe, and I. Sherr. 1986. Strategic planning and the pursuit of reform, economic development and equity. *Journal of the American Planning Association* 52:299-309.

Mishel, L., J. Bernstein, and J. Schmitt. 1997. *The state of working America 1996-97.* Armonk, NY: M. E. Sharpe.

Moss, P., and C. Tilly. 1996a. Informal hiring practices, racial exclusion, and public policy. Paper presented at Politics and Society Conference, November 16, New York.

———. 1996b. "Soft" skills and race: An investigation of Black men's employment problems. *Work and Occupations* 23:252-276.

Newman, K., and C. Lennon. 1995. The job ghetto. *American Prospect* 22:66-67.

Okigaki, A. 1997. *Developing a public policy agenda for jobs.* Washington, DC: Center for Community Change.

O'Regan, R., and M. Conway. 1993. *From the bottom up: Toward a strategy of income and employment generation among the disadvantaged.* Washington, DC: Aspen Institute.

Orr, L. L., H. S. Bloom, S. H. Bell, et al. 1995. *Does training for the disadvantaged work? Evidence from the national JTPA study.* Washington, DC: Urban Institute Press.

Peck, J. 1996. *Work place: The social regulation of labor markets.* New York: Guilford.

Ranney, D., and J. Betancur. 1992. Labor force-based development: A community-oriented approach to targeting job training and industrial development. *Economic Development Quarterly* 6:226-236.

Siegel, B., and P. Kwass. 1996. *Jobs and the poor: Publicly initiated sectoral strategies.* Washington, DC: Aspen Institute.

Steuernagel, B. J. 1995. *The Jobs Gap study: Phase 1, first report of findings*. St. Paul: JOBS NOW Coalition.

Stillman, J. 1994. *Making the connection: Economic development, workforce development, and urban poverty*. New York: Conservation Company.

U.S. Department of Labor. 1991. *What work requires of schools: A SCANS report for America 2000*. Washington, DC: U.S. Department of Labor.

U.S. Department of Labor. 1994. *The role of job training in expanding employment opportunities and increasing the earnings of the disadvantaged*. Washington, DC: U.S. Department of Labor.

Waldinger, R. 1996. *Still the promised city? African-Americans and new immigrants in postindustrial New York*. Cambridge, MA: Harvard University Press.

Waldinger, R. Forthcoming. Black/immigrant competition reassessed: New evidence from Los Angeles. *Sociological Perspectives*.

Weir, M. 1992. *Politics and jobs: The boundaries of employment policy in the United States*. Princeton, NJ: Princeton University Press.

Whiting, B., and J. Taylor. 1997. *Job retention and advancement strategies for low-wage young adult workers*. Boston: Jobs for the Future.

Wilson, W. J. 1996. *When work disappears: The world of the new urban poor*. New York: Knopf.

PART I

BACKGROUND

2

Labor Market Restructuring and Workforce Development

The Changing Dynamics of Earnings, Job Security, and Training Opportunities in the United States

BENNETT HARRISON
MARCUS WEISS

> In tempest time, praise be to *work*; that homeliest, sturdiest, and most reliable of Personal Flotation Devices.
> —*John Barth*

The last national recession in the United States bottomed out during the first 3 months of 1991. For an entire year thereafter, there was no job growth in the country at all, as employers hedged their bets, waiting to see what would happen, meanwhile extracting extra production from existing employees. This was the period that the media called the "jobless recovery." Had it continued, this really *would* have been a new development in the history of U.S. labor markets.

AUTHORS' NOTE: An earlier version of this chapter appears in Bennett Harrison and Marcus Weiss, *Workforce Development Networks: Community-Based Organizations and Regional Alliances* (Sage Publications, 1998).

But by the spring of 1992, jobs were again being created, and at just about the average rate that had prevailed among all eight previous business cycle recoveries going back to the end of World War II. Even including that year of "jobless growth," the U.S. economy created more than 11 million new jobs between spring 1991 and fall 1996.

Contrary to many casual impressions—and not a few analyses by the experts and in journalistic commentaries—the U.S. Bureau of Labor Statistics (BLS) makes it quite clear that, if both new jobs created and hires to replace vacancies in "old" jobs are counted, relatively low-skilled positions (defined by the BLS as "jobs that can be learned quickly and that generally do not require post-secondary education") accounted for most of the wage and salary employment created in the United States between 1983 and 1993 (Rosenthal 1995). Moreover, projections based on the BLS's widely used *Occupational Outlook Handbook* conclude that, for the period 1992-2005, again considering both new jobs and replacement vacancies, only about one out of eight higher-than-average growth occupations will require a college degree, and fully two-thirds will call for no more than a high school diploma (S. Mangum 1995). It really is quite amazing how powerfully the myth of a steadily disappearing demand for low-skilled labor in the United States has been perpetuated.

None of this should be taken to imply that education is unimportant—that would be foolish. Nor is it inconsistent with Harvard University economist Ronald Ferguson's (1996) finding that racial differences in basic math and language skills (as measured by standard written tests) are strongly correlated with the racial wage gap. But it does say that the alleged disappearance of low-skilled job opportunities in the United States has been exaggerated. There is, and will continue to be, considerable room in the economy for workers with modest formal schooling. The main problem is to improve dramatically the quality and reliability of *basic* skills for the majority of youth who will not go beyond high school. They, along with adults undergoing retraining, are increasingly expected by employers to be better equipped to learn new skills on the job, to take further training, and to be, by mainstream standards, willing and able to accept the disciplinary requirements of most workplaces.

In an institutionally racist society, of course, the continued creation of jobs with modest skill requirements does not guarantee equal access of all prospective workers to even these "low-end" jobs. In the big cities especially, a new wave of ethnographic scholarship—the scholarship of direct engagement, not relying solely on the manipulation of census data—has documented complex and intense competition for these jobs, between men and women, between residents and newcomers, and, among the latter, among groups that have traversed very different paths (experienced different modes of "insertion") into the urban economy (e.g., Morales and Bonilla 1993; Sassen 1989; Thompson 1997; Waquant 1994; Waldinger 1996). In this competition for low-end jobs, racial divisions and years of residence in the city go far toward explaining who holds which places in the system.

The upshot is that the signal problems that plague urban labor markets—and employment systems everywhere—consist mainly of low wages, job insecurity, and inadequate opportunities for on-the-job learning, *not* some technologically (or otherwise) driven "twist" toward employers' greater need for high skills per se. Barbara Bergmann's old "crowding" hypothesis, in the context of greatly weakened labor regulation (enforcement of national wage and hour and safety and health standards), seems a far more useful way to think about what is happening than the hopelessly technologically determinist myths about labor market restructuring that have come to dominate mainstream discourse, especially within academic economics.[1]

Skills, Earnings, Race, and Sectoral Shifts in the Urban Economy

The categories *skill, schooling,* and *ability to learn* require careful, nuanced distinctions. A new survey of nearly 3,000 employers in the four metro areas of Los Angeles, Detroit, Boston, and Atlanta, conducted by Harry Holzer, reflects the importance of recognizing these distinctions. Holzer (1996) reports that the extent to which employers require high school diplomas, specific experience, and prior training and references—*but not postsecondary education*—as conditions for hiring into entry-level jobs that entail tasks such as daily reading and writing, arithmetic, use of computers, or dealing with customers increases the probabilities that Blacks and women will be hired (vis-à-vis White men), as well as the level of their starting hourly wage rates.[2]

A recent national survey of hiring, training, and management practices in a representative sample of (again, by coincidence) 3,000 companies—the first official survey of its kind—was conducted by the U.S. Census Bureau in August and September 1994 for the National Center on the Educational Quality of the Workforce and the Institute for Research on Higher Education, both based at the University of Pennsylvania. The survey results show that employers place higher priority on "attitude," "communication skills," "previous work experience,"

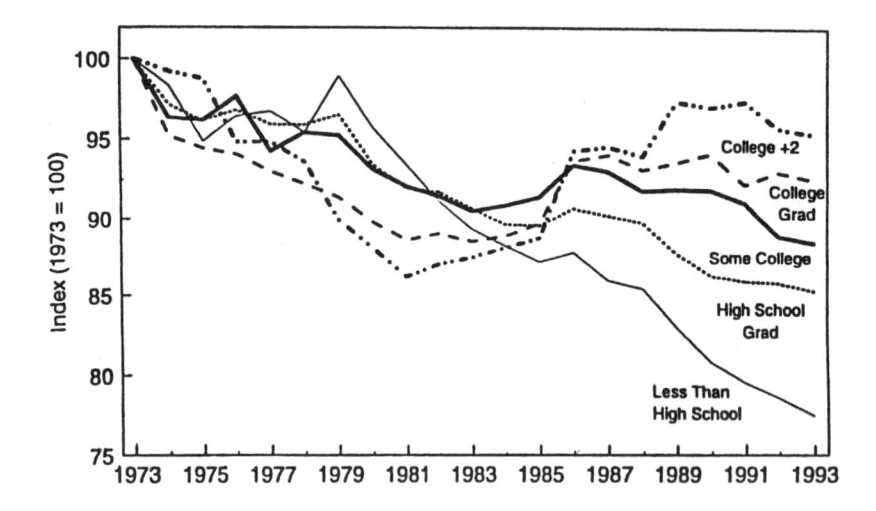

Figure 2.1. Hourly Wages by Education, 1973-1993
SOURCE: Mishel and Bernstein (1994, 141).

"recommendations from current employees," "recommendation from previous employer," and "industry-based credentials certifying skills" than they do on years of schooling, test scores, grades, or teacher recommendations.

At the same time, managers in small focus groups told Robert Zemsky, the center's codirector and the institute's director, that they were extremely skeptical of the quality of most young graduates of the nation's schools (Applebome 1995). Employers say they want people with *better, more reliable* schooling in job-relevant skills, but not necessarily people with *more* schooling per se (see Zemsky 1994).[3]

Thus, the economy continues to generate job opportunities, both through the creation of new employment and through turnover in existing slots. And employers are demanding new workers with better basic skills but hardly with rocket science under their belts (or in their heads—the typical level of "computer literacy" that employers demand can often be taught in one solid semester in a high school or community college computer lab). The problem is that average wages and benefits have continued to decline, even though average productivity and profits are growing. This is by now a well-researched subject, with substantial agreement on the "facts" (if not on the relative importance of the causes). Figure 2.1 displays the pattern of inflation-adjusted hourly wages since 1973,

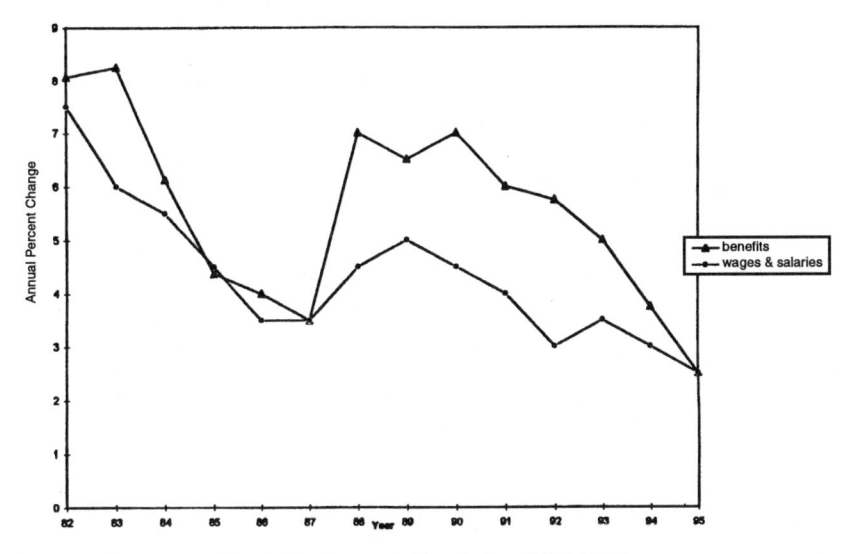

Figure 2.2. Nominal Hourly Employment Cost Index, 1982-1995
SOURCE: U.S. Bureau of Labor Statistics (reprinted in Hershey 1995, 2).

by education. Clearly it is those with the fewest years of schooling—precisely the segment of the population that is disproportionately concentrated in the inner city and in poor rural areas—who have lost the most ground. But during the 1990s, even those with 4-year college degrees have seen their real earnings growth flatten (Mishel and Bernstein 1994, 140-145, 364-366). Women working full-time experienced rising median real weekly earnings over most of the past 20 years (even as the average earnings of men fell steadily), but that trend came to an abrupt end in 1994. Since then, women's earnings have been falling too (Mishel 1995, 61).

The government keeps statistics from the vantage point of employers, as well. The BLS's "employment cost index" measures spending by companies, government agencies, and nonprofit organizations on pay and benefits for each hour worked. Figure 2.2 graphs percentage changes between each quarter of a year and the corresponding quarter a year earlier. All civilian employees are included. "Benefits" include employer expenses for health insurance, vacations, holidays, sick leave, shift premiums, life insurance, pensions, unemployment insurance, and severance pay. The annual growth rates of both wage and benefit costs to employers per hour of work show strong declining trends since at least the early

TABLE 2.1 Median Annual Earnings of Year-Round, Full-Time (YRFT) U.S. Workers: Comparing Public Employees to All Workers, by Race and Sex, 1989

	Government Sector		All Sectors	
	Median Annual Earnings ($)	% Working YRFT	Median Annual Earnings ($)	% Working YRFT
Men				
Black	26,626	80	21,971	64
all races	31,274	85	26,269	66
Women				
Black	21,346	73	18,036	57
all races	21,760	75	18,351	52

SOURCES: Figures for Blacks are from U.S. Bureau of the Census (1993, Tables 46-47) and refer to earnings in 1989. Figures for all races are from U.S. Bureau of the Census (1992, Table 33), deflated back to 1989 for purposes of comparison, using the CPI-U change of 12% inflation.

1980s (with the predictable cyclical bump up associated with tighter labor markets, prior to the 1990-1991 recession). These figures are not adjusted for inflation. After increases in the cost of living are accounted for, wage and salary growth by 1993 was actually falling *behind* changes in the cost of living (Hershey 1995).

Of course, what to managers appears as "success" in containing the growth of contributions to their employees' health insurance premiums and other benefits spells growing hardship for a rising share of the population, in a society in which the already inadequate contribution of the government to social insurance—especially to welfare for poor relief—was effectively compromised in 1996 and is likely to erode further in the years immediately ahead.

As for urban communities of color, for many years it was true that African Americans, in particular, depended on and benefited from a healthy, growing public sector for jobs and for social services of all kinds.[4] Black men and women employed in government jobs average significantly higher wages than do African Americans who are employed in the private sector (see Table 2.1). A much larger fraction of them are likely to be employed year-round and full-time, as well. The advantage of working for government is especially great for women. Actually, what the Census Bureau data displayed in Table 2.1 reveal is that public sector employment is the great equalizer in American working life. For Blacks and other workers of color, the likely decline in government's long-standing one-sixth share of total employment in the country under any of the "balanced budget" austerity regimes now being debated in Washington is bad news, indeed.

A second issue concerns the current wave of mergers and consolidations among banks, hospitals, and other typically urbanized service sector firms. These are the employers who have been the least likely to shut down existing operations to relocate to the suburbs (let alone overseas), choosing instead to set up additional branches. They, too, have for many years been important sources of employment opportunities for workers of color ("health," in particular, shows up in everyone's sectoral analyses of "good bets" for targeted economic development and job training programs). They will continue to be important—especially the health sector, given the graying of the population. Moreover, as the big hospitals continue to downsize and outsource services to smaller independent contractors, nursing care facilities, and other organizations, under the cost-cutting aegis of "managed care," new jobs will be created in these "outside" workplaces. Still, the extent of the growth of this sector as a whole will depend in part on what happens to our political decisions about public spending, regulation, and the pace of deregulation in the years ahead. And whatever the extent of the slowdown, the burden is likely to fall disproportionately on the residents of low-income communities of color, in terms of both services forgone and jobs and wages lost.

To sum up: Job opportunities *are* being created, and the great majority of them do *not* (and, in the years just ahead, probably will not) require a college education. On the other hand, those sectors that have been especially important sources of employment for the residents of largely Black, Latino, and other urban neighborhoods—hospitals and other health facilities, government itself, and, more recently, banks—are no longer creating jobs at a pace commensurate with their track records of even the recent past. Finally, average wages and even benefits or, by another measure, their rates of growth continue to decline, year after year, for a large fraction—by some accounts, perhaps four-fifths—of the population.[5]

Declining Job Security and Diminishing Returns to Experience and Seniority: The Devolution of Internal Labor Markets

Far-ranging changes are under way in how companies organize work: how they produce, where they locate the work to be done, and whom they hire to do the work. A central feature of these changes is what now appears to be a long-run ("secular" as opposed to periodic or "cyclical") move away from what, at least for the most profitable operations of the biggest, most visible, and most influential firms and agencies, constituted the dominant, if never universal, employment system of the post-World War II era.

In the three decades or so after 1945, with job tasks substantially standardized and broken down into a fine-grained division of labor, managers commonly hired in new young workers at the bottom of the job ladder, trained them on the job, promoted within the organization, and paid wages more in accord with seniority and experience than with individual productivity or current firm performance. Although few companies ever formally committed themselves to literal "lifetime" employment, there was widespread expectation that, at least for White men, but gradually for others as well, there would be a high degree of long-term job security, with occasional disruptions in work, triggering the receipt of unemployment insurance and other forms of temporary income maintenance. Seniority systems were promoted by unions, and within the civil service, as a means of achieving fairness. Business came to support these systems because, by according greater job security to more experienced employees, they could better count on these individuals to train and otherwise teach skills and know-how to younger workers within the organization. The question of "who is poor" or underemployed then amounted importantly to understanding what tracking mechanisms in the society systematically blocked which groups from having access to jobs in these "internal labor markets."[6]

Whether there is a well-defined, coherent new "system" of work coming into existence—accommodating to globalization, deregulation, the shortened shelf life of products, the enhanced capability of more and more competitors to erode quickly the advantages of (play catch-up with) innovators through "reverse engineering," and generally heightened competition—is still an open question. But there is a growing consensus among scholars that the old system is coming apart. This is not the place to examine all of the many aspects of this "devolution" of internal labor markets, the growth of "flexible" forms of work organization, and the implications for training and economic development.[7] But we can at least sketch the main features of what a number of researchers believe is happening to how work and labor markets are coming to be reorganized.[8]

The changes are motivated by a complex of factors: deregulation, greater actual or potential competition from abroad, growing numbers of corporate hostile takeover attempts, and other signals from stockholders that put a premium on short-term firm performance. All of this has made managers increasingly conscious of short-term fixed (or, as economists like to describe the wages of workers on long-term implicit contracts, "quasi-fixed") costs and committed to reducing them whenever and as much as possible. First with IBM, then with

Xerox, and most recently with AT&T and the big banks, the stock market has instantly rewarded companies that cut costs through consolidations, mass layoffs, and wage and benefit rollbacks by bidding up share values, which only further encourages this sort of management behavior. Even as progressive voices—from former U.S. Secretaries of Labor Ray Marshall and Robert Reich to labor economist Richard Freeman and political theorist and activist Joel Rogers—were advocating that more companies should take (and stick to) the "high road" in labor policy, the business community has been organizing itself (e.g., through the Labor Policy Association) to advocate for, and legitimate, greater "flexibility" and management discretion in work arrangements.

What this amounts to, in practice, is a proliferation of different forms of work organization, blurring the traditional distinctions between "core" and "periphery," "permanent" and "contingent," and "inside" and "outside" employees and between "primary" and "secondary" labor markets. Thus, managers employ some workers in more or less routine wage and salary positions inside the firm or agency, but they also pay temporary help agencies, management consulting firms, and other contractors to provide employees—ranging from specialized computer programmers to janitors and clerical personnel—to work alongside the "regular" people but on short-term assignments and under the management of the contractor. Companies, colleges, and hospitals outsource work formerly performed in-house to outside suppliers, here and abroad. They also shift some work from full-time to part-time schedules, in part to avoid federal labor regulations covering wages and benefits that have been interpreted by the courts as not covering "leased" workers. "Temp" agencies and other contractors are being used increasingly by managers in "focal" firms as mechanisms for screening potential regular employees, with candidates serving their "probation" on the payrolls of the contractors before moving into the focal companies or agencies. This creates the possibility for further inequities, because two persons working side by side for an outside contractor may be equally competent, yet one will eventually be absorbed into a full-time job while the other remains bouncing from one temporary assignment to another.

Two concrete expressions of this growing heterogeneity in work organization and practices, both among and within particular employers, are declining employment security and more uncertain wage and salary prospects over time. Surveys conducted by the American Management Association show that managers are increasingly regarding layoffs as "strategic or structural in nature," rather than as responses to short-term, temporary business conditions (cited in Cappelli 1995, 577).[9] The fraction of laid-off workers who can expect recall into their old jobs has been lower in the post-1991 recovery than in the previous four

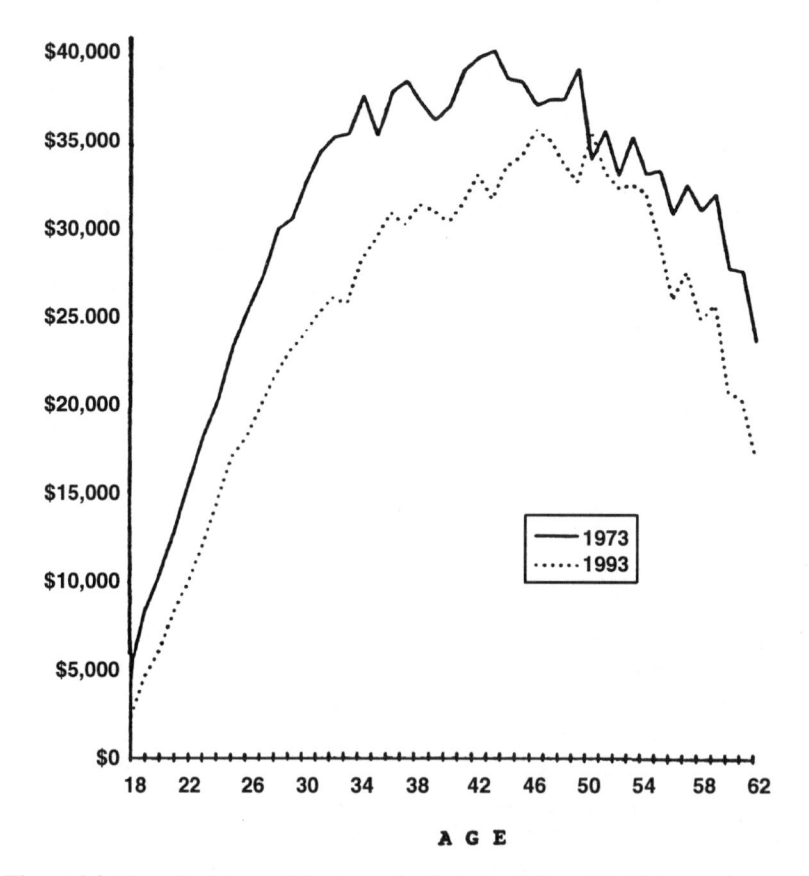

Figure 2.3. Mean Real Annual Earnings of All 18- to 62-Year-Old Males, by Age, 1973 and 1993 (in 1993 dollars)
SOURCE: Pines et al. (1995, 17).

national business cycles, and those who do get reemployed somewhere are more likely than before to land in part-time jobs (Cappelli 1995, 578).

The flatter organizational structure—yet another workplace "innovation," the introduction of which in recent years has been responsible for many of the layoffs of middle-level managers—contributes to reduced promotion opportunities within the surviving internal labor markets of companies and agencies. More generally, the payoff to seniority, as measured by age-earnings profiles, is shrinking over time, both across the workforce (Figure 2.3) and, more precisely, over the careers of specific individuals, including those who stay with one employer.[10] Also, compared with expectations formed in earlier periods, such as the early 1970s, or with the average earnings profiles for their age groups, workers in the 1980s and 1990s face increasingly fluctuating earnings from one year to the next.[11] Quite apart from questions of "fairness," Cappelli and others are questioning the long-run implications of this growing instability in earnings streams for the continued smooth functioning of a consumer economy built on households borrowing against future income and having to make small regular payments over time for everything from housing and cars to their children's education.

In sum, the mix of types of jobs and work schedules is becoming more diverse. In Cappelli's careful and sober formulation: "There is a continuum between 'pure' internalized arrangements and complete market determination along which employers are moving, and the argument here is that, on average, practices are shifting along that continuum," from the former in the direction of the latter (1995, 570). If this were mostly a voluntary development, serving the needs of a greater share of the population, that would be one thing. But all serious researchers agree that these changes are being initiated mainly on the "demand side" of the labor market, with managers seeking to reduce their exposure to long-term fixed obligations. It appears that a growing fraction of pay is becoming "contingent"—on individual job performance, on the fortunes of the employer, on the current state of the "animal spirits" in the stock market, or on what the company thinks it can get for its money by turning to suppliers in other, lower-cost locations.

Implications for Employment Training and Workforce Development

Although wages, benefits, and opportunities for secure long-term employment in U.S. labor markets are all being buffeted by forces far beyond the control of any individual job seeker, we have seen that new jobs and job openings *are* being created, and at about the average pace of the last half century. The pace of job creation continues to be inadequate to provide employment for all who seek it. But jobs are being created—in the millions. That in itself would make it compelling for community-based organizations to have an interest in access to high-quality employment training for the residents of their neighborhoods.

In fact, the case for such interest is even stronger. This is because workforce development—the constellation of activities, from recruiting, placement, and mentoring to follow-up, of which the actual training is but one element—involves not just the "production" of skills, but also the enhancement of trainees' ability to learn and their socialization to working around others in settings defined by managers in private firms and public or nonprofit agencies. Employers select job applicants based on their (the employers') perceptions of all of these. Indeed, for entry-level jobs, Ronald Mincy, Chris Tilly, and others have argued that "skill," in the narrow, technical sense, may be less important to many gatekeepers than perceived attitudes and socialization—what have come to be known as the "soft" skills (Pouncy and Mincy 1995; Moss and Tilly 1996b). This has implications for hiring discrimination, to the extent that—as Tilly and others believe—employers are more easily able (and likely) to discriminate against Black men when screening and hiring are done informally and subjectively. The soft skills are surely more likely to be "processed" informally than through actual tests or by certification (Moss and Tilly 1996a). By this reasoning, it is probably becoming easier for employers to screen out Black men whom they do not want to hire for reasons other than lack of skill per se.

Obviously, training without economic development will always leave some people unemployed or underemployed at poverty wages; in the game of musical chairs, everyone doesn't get to sit down if there aren't enough seats. But for any given supply or rate of growth of job vacancies (i.e., demand for labor), those who have more (and higher-quality, or more unambiguously certified) training tend to be the ones most likely to get a seat. Scholars differ among themselves as to how much this is the result of individuals' getting "smarter" and how much may be due to the reputation and "connectedness" of the training organization. But there is considerable agreement within the research and evaluation communities that there exist at least some training programs that do (for at least some populations) show positive private and social payoffs.

What Kinds of Training Work Best and for Whom?

The most recent bevy of evaluations, surveys, and metasurveys on the costs and benefits of employment training in particular, and of workforce development more generally, substantially replicate findings that have been well-known to researchers and policy makers in this field for many years. In other words, as far as the question of "what's working (and what's not)" goes (as it is put in the title of a 1995 report from the U.S. Department of Labor), surprisingly little has changed over time.

The most general conclusion to be drawn from this substantial literature is that for adults, there are "serious doubts about the effectiveness of stand-alone basic education and GED programs" (Savner 1996, 5). Cornell Professor John

Bishop, former DOL Chief Economist Lisa Lynch, MIT's Paul Osterman, and other specialists have amply demonstrated that employer-centered training produces the greatest payoffs (see, e.g., Bishop 1994a, 1994b; Knocke and Kalleberg 1994; Osterman and Batt 1993). Indeed, Bishop has shown that employers themselves clearly benefit from such investments, through increased worker productivity and reduced turnover. Training sponsored or conducted directly by employers generates relatively greater benefits, in terms of wage improvements and reduced chances of early subsequent unemployment, than does any other form of training. Vouchers provided to adults to return to school, so that they can then go on to search for jobs on their own, have proven to be especially ineffective.

The earnings of young people still in school—especially young women of all races and ethnicities—*are* on average enhanced by occupational training in high schools, community colleges, and (some) vocational-technical organizations. By contrast, the payoff to school-based training for young African American men in recent years has been essentially zero (this is *not* true for armed forces training, the universally acknowledged effectiveness of which may provide a clue to what's going on, one to which we will return later).

The group for which practically none of the existing approaches has worked is out-of-school youth, with a handful of exceptions, notably the Job Corps and the San Jose-based, largely Chicano Center for Employment Training (CET), a program we characterize as the single most innovative, and demonstrably successful, nongovernmental training apparatus for poor people in the United States. Short-term (3- to 6-month) classroom training is generally not very effective for any group, with CET being the remarkable exception to the rule (see Giloth, Chapter 1, this volume).

Continuing Systematic Underinvestment in Young People by the Private Sector

Company-based (or closely linked) training may show the best results, but— especially for entry- and other lower-level, generally younger employees—it is hard to find. Although precise estimates vary, the general conclusion that U.S. firms systematically underinvest in firm-specific training, especially in their less-skilled employees, appears in study after study. According to Lynch (1993), the U.S. private sector underinvests in firm-specific training compared with private firms in Japan, Australia, France, Germany, the Netherlands, Norway, Sweden, and even the United Kingdom.[12]

As for the bias against spending on youth (and on low-skilled workers generally), Ray Marshall and Marc Tucker (1992) claim that at least two-thirds of the annual spending of all U.S. companies on education and training during the 1980s was invested in their college-educated professional employees. Since the early 1980s, only 4% of non-college students aged 16-25 have received formal company-centered training of at least 4 weeks' duration (Lynch 1993). Between 1983 and 1991, the average duration of formal training paid for and provided by employers for learning a new job declined substantially, especially for workers with fewer than 10 years of seniority (cited in Cappelli 1995, 577). Moreover, although recent surveys show that the majority of private firms do make some investments in worker training, the private sector American Society for Training and Development reports that a tiny handful are responsible for the vast majority of the total investment—one-half of 1% of all firms spend 90% of the total (cited in Henkoff 1993, 62).

The traditional explanation for this investment behavior on the part of private companies, in the context of American political institutions, is that workers here are more likely than those elsewhere to take whatever training they receive inside a particular firm and "jump ship" to another. At the level of high theory, Nobel economist Gary Becker many years ago drew a powerful distinction between firm-specific and more "general" training, predicting that private businesses would always underinvest in the latter because it could be used by footloose employees ultimately to benefit competitors. Bishop (1994b, 25) confirms that, over the past 20 years, a smaller fraction of American workers had been on their current jobs (with their current employers) for more than 5 years than almost anywhere else; the rates for Japan and Germany were half again as great. This nexus forms the basis for economists' widely shared classification of general ("portable") training as a "public good" that, if not subsidized by government (however and in whatever settings it is actually provided), will be systematically underproduced.[13]

At an empirical level, part of the explanation for the relatively low average flow of private spending on employer-centered training is that most firms are small, and the incidence of spending on training declines dramatically as the size of the establishment decreases (Bishop 1994b, 8).[14] The reasons are apparent: Small firms have smaller budgets, less organizational slack for managing such activities, and—partly because they also, on average, pay lower wages and benefits—confront higher probabilities that employees will "job hop" sooner. As with medical care, occupational health and safety, environmental concerns, and access to information about new technologies, it is precisely the smaller business sector of the economy that is most in need of a combination of the carrot of targeted public policies and the stick of regulatory sanctions if we are to maintain a higher overall standard of living.

There are, of course, numerous examples of private companies (generally among the largest ones) that have made major commitments to firm-based training. Motorola, Xerox, Corning, General Electric, Hewlett-Packard, Ford, Federal Express, AT&T, and Siemens are among the companies most frequently cited for their relatively expansive approaches to employee training (see Henkoff 1993). Whether these "success stories" add up to a trend toward restructuring corporate environments as "high-performance workplaces" is, however, still contested. Even many of the early advocates, such as Appelbaum, Bluestone, and Cappelli, must be said to have become "cautiously pessimistic."

Unions have also (and traditionally) played an important role in negotiating to get employers to make investments in training that they would not otherwise have made. A good example is afforded by AT&T. Under a contract dating back to 1986, the firm, the Communications Workers of America, and the International Brotherhood of Electrical Workers established a joint labor-management training fund to support training of workers over and above what the company normally provides, targeted especially to help employees meet the entry-level requirements for changing jobs within the firm or for preparing for new employment outside AT&T altogether. Similar funds exist in the automobile industry, notably between General Motors and the United Auto Workers. The durability of the AT&T commitment is currently being tested as never before, given the mass restructuring-related layoffs now under way in that company.

Prospects for Partnerships Between Private Companies and Community Colleges, and Implications for Job Training in Low-Income Communities of Color

Private companies have always engaged outside organizations to provide training services for and with them. In any city in the United States there are dozens, even hundreds, of private and not-for-profit vendors providing all manner of training services on contract to businesses, either directly or through one or another federal or state-subsidized program. The major federal government programs remain those created under the Job Training Partnership Act, the welfare-to-work activities mandated by the Job Opportunities and Basic Skills program under the 1988 Family Support Act, and the programs created under the Carl D. Perkins Vocational and Applied Technology Act of 1990 and, most recently, the 1994 School to Work Opportunities Act, aimed at integrating school-based and workplace-connected ("concurrent" or "contextual") learning. There are also many state-initiated programs that entail cooperation with private industry. All of this legislation is currently being reconsidered in ongoing negotiations between Congress and the White House.

Across all of these many activities, much hope is being placed in the emergence of 2-year community colleges as crucial sources of customized training for firms and as important bridges joining the streets, high schools, employers, and 4-year colleges and universities. The performance of community colleges has inevitably been uneven. There is, moreover, a lingering tension between these institutions' "vocational" training mission and those professional educators who emphasize the role of community colleges as that of providing remedial or transitional education for students on their way to "regular" 4-year colleges (Harrison, Weiss, and Gant 1995, 28-34). Elijah Anderson's long-standing concern about the negative consequences of a mismatch between the races of the mostly White instructors and young African American trainees in the job training programs of the 1960s and 1970s (especially the difficulty of developing sufficient mutual trust to support effective mentoring) applies to the community colleges as well (see Anderson 1990).

These concerns notwithstanding, there are a growing number of well-documented cases of community colleges working closely with companies.[15] This has not gone unnoticed by community-based organizations, which are making efforts in several places to partner more closely with these key institutions that mediate between citizens and employers. At the same time, a growing number of companies are providing advice, instructors, and equipment to community colleges. They are jointly developing curricula and becoming acquainted with students and trainees long before the latter complete their courses, through instruction, summer, and "co-op" jobs. In some places, 4-year colleges and universities are making themselves key players in these citizen-neighborhood-college-employer networks too (Figure 2.4) (Fitzgerald 1995, 34).

The Time Is Right for New Approaches

It cannot be stressed enough that even the most effective, well-targeted workforce development programs will not, by themselves, solve the problem of insufficient job creation at what sociologist Lee Rainwater used to call "get along" income levels. It has taken mainstream economists a long time to come around to acknowledging the existence, let alone the centrality, of dysfunction on the "demand side" of the labor market. Policies and programs that fail to

```
                    Employers (Businesses, Agencies)
                              ↑  ↑  ↑↓
                    Business,
                    Outreach,
                    Manufacturing,   Instructors,   Train and
                    Modernization,   Equipment,     channel
                    Industrial       Curriculum     residents into
                    Extension        Design         employers

Membership in           COMMUNITY                        4-Year college
networks        ⇌       COLLEGES AS              →
of community            INTERMEDIARIES                   student transferees
colleges,
some universities
                              ↑              ↑
                    EXAMPLES:

                    Pittsburgh Partnership for
                    Neighborhood Development and
                    Community College of Allegheny
                    County                              Residents of
                    QUEST and Alamo Community           low-income
                    College (San Antonio)               communities
                    New Community Corp. (Newark)
                    and Essex County Community College
                    HOPE and Wayne County
                    Community College (Detroit)

                              Community-based
                              organizations
```

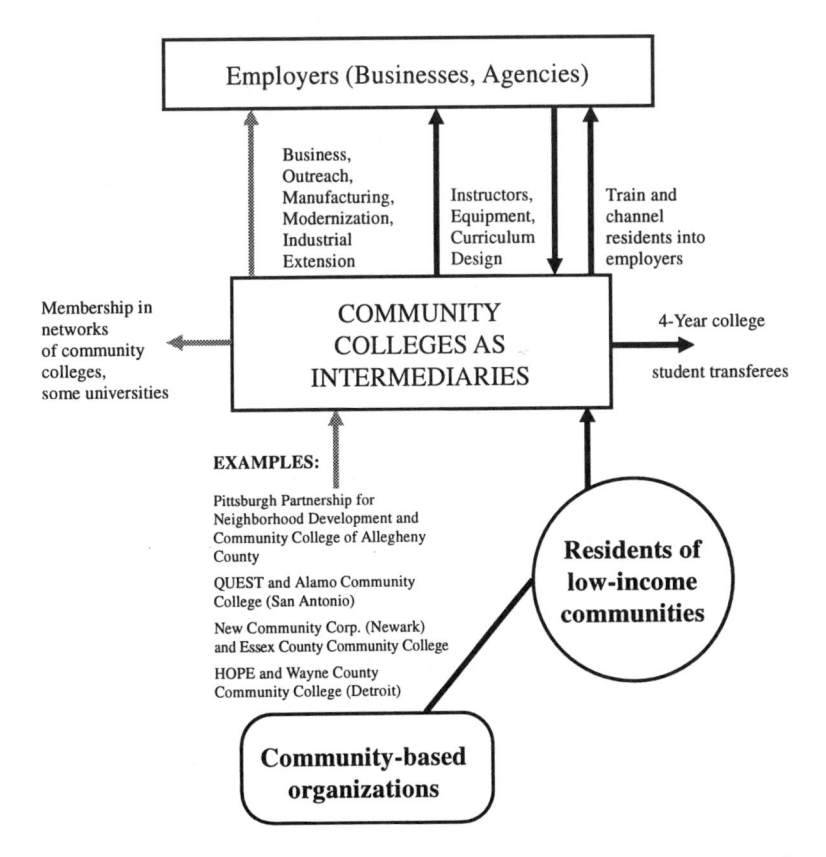

Figure 2.4. Community Colleges as Intermediaries in Workforce Development and Industrial Modernization Programs

address the need for economic development—within inner-city neighborhoods, throughout cities, and across whole metropolitan regions—are destined to leave people with dashed hopes and continued frustrations. On the other hand, in the context of comprehensive economic development activities, the "supply side" of the labor market—education, training, and job search—comes into its own.

To be sure, mainstream workforce development/employment training programs have never had good press—at least partly because companies and the public have never been fully committed to them. Except that the design and management of these programs provided for the "empowerment" of a quite substantial cadre of now middle-class professionals of color, at all levels of government and in the private and foundation sectors, it is remarkable how little has changed since the early 1960s. With only slight exaggeration (and the occasional error of omission), veteran employment training expert and administrator Garth Mangum summed up 33 years of experience with federal workforce development policy this way in a 1995 lecture to congressional staffers:

> Benefits have [generally] exceeded costs for adult programs, but always by narrow margins. . . . No major youth program except Job Corps has ever had a favorable ratio of benefits to costs. . . . Program proliferation has always been an irritant, interfering more [however] with orderly administration than with program outcomes. . . . The relative roles of federal, state and local governments have constantly fluctuated. . . . The mix between preventative preparation and second chance remediation has always been an issue, as has been an antipoverty vs. mainstream [targeted vs. universalist] emphasis. . . . The same can be said for welfare reform. Employed self-sufficiency for AFDC parents has long been a lodestone. Few have doubted that it was obtainable at the price of a combination of remedial education and skill training, child care and health care, subsidized private and public service employment and probable continued income supplementation for many. But so far no one has been prepared to pay the price. (1995, 4-6)

Perhaps the mainstream programs can be fixed. Perhaps the currently bleak political climate in Washington will not, after all, undermine the successful working out of the bugs in the latest employment training reform movement, "school to work," with its dreams of concurrent learning (academic education *plus* real-world work experience), "one-stop shopping," and national educational standards.

In the meantime, there is surely no reason to hold back on learning more about, and promoting the fortunes of, more *unconventional* approaches to workforce development, especially those that centrally involve low-income community-based organizations and activists of color. The essays that follow in this book describe a number of such approaches. Can they make a difference? Let's hope so. Are they desperately needed in a time of profound structural change and urban crisis? You bet.

Notes

1. For the latest application of the "crowding hypothesis" that she first developed in the early 1970s, see Bergmann (1996).

2. However, Holzer also believes that formal testing by employers generally works against Black men, because they are less likely to pass the tests. Chris Tilly and his colleagues think that, on balance, formal screening and hiring procedures (although not necessarily testing per se) help Black men, whereas informal methods (which are probably on the rise, with the reduced enforcement

of affirmative action and related regulations by government) allow more room for employers to discriminate arbitrarily (Moss and Tilly 1996a; Kirschenman and Neckerman 1991).

3. Mainstream approaches to the apparently rising "returns to skill" are abundant. One of the best is described by Freeman and Katz (1994). For some thoughtful alternative interpretations of the same numbers, see Howell (1994, 1997), Levy and Murnane (1992), and Mishel and Teixera (1991).

4. The most comprehensive statistical documentation and thorough literature review on the jobs aspect of this assertion is contained in a recent working paper by Michael Leo Owens (1996).

5. By late 1996, under the pressure of what, these days, count as relatively low rates of unemployment, there were signs of small increases in real average hourly wages (although even this claim was contested; what was happening to average wages depended on which data sources one consulted). On the other hand, no improvement in the now two-decade-long increase in earnings *inequality* is in sight. On changes in the distributions of family incomes, wealth, individual labor market earnings and benefits, the most comprehensive survey is by Frank Levy (1995). For especially accessible treatments of particular aspects of what Lester Thurow once called "the surge in inequality," see Bluestone (1995), Krugman (1992), and Wolff (1995).

6. For a sampling of classic and enriched views of the postwar system of internal labor markets, and their implications for employment training, racial discrimination, and urban poverty, see Berger and Piore (1980); Doeringer and Piore (1985); Gordon (1971); Gordon, Edwards, and Reich (1982); Harrison (1972, especially chap. 5); Jacoby (1985); Osterman (1984, 1988); and Peck (1996).

7. For in-depth explorations of these questions, see Appelbaum and Batt (1994), Bailey and Bernhardt (1996), Cappelli (1995), Cappelli et al. (1997), Harrison (1994), Osterman (1994, 1995), Pfeffer and Baron (1988), and Piore and Sabel (1984).

8. The following material draws especially heavily on the recent writings of Peter Cappelli. There, and elsewhere, readers can also learn about other new developments, from the growth of teams to the breakdown of narrow job descriptions and the enlargement of many existing jobs, with those who do have work being required/enabled to perform a greater variety of tasks, often using productivity-enhanced new technology. The changes that are occurring in the organization of work in the United States are not all "bad"—provided one can break into the system.

9. Such permanent layoffs ("displacements") are currently actually higher for managers than for those in other positions, after other characteristics are taken into account.

10. The flattening of age-earnings profiles reflected in Figure 2.3 is suggestive, but longitudinal studies on specific workers and on cohorts are more definitive. For such evidence, see Marcotte (1994), Bluestone and Rose (1997), and Rose (1995). A new study from the Employee Benefit Research Institute reports that, between 1991 and 1996, median job tenure for men aged 25-64 fell by nearly 20% (see "Economic Trends," 1997).

11. Recent technical studies include those conducted by Bailey and Bernhardt (1996) and Gottschalk and Moffitt (1994). Some of this increasing volatility is due to the growing incidence of performance-based pay and bonuses and, as such, is built into the new work systems by design. Some is also probably attributable to declining job tenure, although (as Cappelli 1995 explains) "tenure" is a confusing indicator of job security, because it is affected by changes in both quits (mostly voluntary) and layoffs (usually involuntary). Unhappily, the U.S. Census Bureau stopped collecting data distinguishing between quits and layoffs back in 1981, at the beginning of the Reagan administration.

12. Bishop adds, "American employers appear to devote less time and resources to the training of entry-level blue collar, clerical and service employees than employers in Germany and Japan" (1994b, 24).

13. American employers do contribute part of the social cost of general training, by forgoing the extra profits associated with putting up with generally less productive new workers. But then they are also partially compensated by being allowed—by social custom and by union contracts—to pay lower wage rates to apprentices and other novices.

14. The BLS confirmed this well-known relationship in a recent survey of formal training programs in the private sector (see Bureau of National Affairs 1994). Of course, the proportion of private businesses that are small is much greater in Japan and Germany than in the United States (Harrison 1994, chap. 2). Yet Japanese and German firms still manage to invest considerably more in worker training than do their American counterparts. Culture, social norms, and national public policy really do make a difference.

15. Evaluation reports and meta-analytic studies that are especially sensitive to the place of community-based organizations in the community college-private employer nexus include the work of Fitzgerald (1995), Fitzgerald and Jenkins (1997), Rosenfeld (1995), and Rosenfeld and Kingslow (1995).

References

Anderson, E. 1990. Racial tension and job training. In *The nature of work: Sociological perspectives*, edited by K. Erikson and S. P. Vallas. New Haven, CT: Yale University Press.

Appelbaum, E., and R. Batt. 1994. *The new American workplace.* Ithaca, NY: Industrial and Labor Relations Press.

Applebome, P. 1995. Employers wary of school system. *New York Times,* February 20, 1995, A1.

Bailey, T. R., and A. D. Bernhardt. 1996. *In search of the high road in a low-wage industry.* Working Paper 2. New York: Institute on Education and the Economy, Teachers College, Columbia University.

Berger, S., and M. J. Piore. 1980. *Dualism and discontinuity in industrial societies.* New York: Cambridge University Press.

Bergmann, B. 1996. *In defense of affirmative action.* New York: Basic Books.

Bishop, J. 1994a. The impact of previous training on productivity and wages. In *Training and the private sector: The international comparisons,* edited by L. M. Lynch. Chicago: University of Chicago Press.

———. 1994b. *The incidence of and payoff to employer training.* Working Paper 94-17. Ithaca, NY: Center for Advanced Human Resource Studies, School of Industrial and Labor Relations, Cornell University.

Bluestone, B. 1995. The inequality express. *American Prospect* 20:81-93.

Bluestone, B., and S. Rose. 1997. Overworked *and* underemployed: Unraveling an economic enigma. *American Prospect* 31(March-April): 58-69.

Bureau of National Affairs, Inc. 1994. [Results of a survey of formal training programs in the private sector]. *Daily Labor Reporter,* September 26, B-19.

Cappelli, P. 1995. Rethinking employment. *British Journal of Industrial Relations* 33:563-602.

Cappelli, P., L. Bassi, H. Katz, D. Knoke, P. Osterman, and M. Useem. 1997. *Change at work.* New York: Oxford University Press.

Doeringer, P. B., and M. J. Piore. 1985. *Internal labor markets and manpower analysis.* Armonk, NY: M. E. Sharpe.

Economic trends. 1997. *Business Week,* January 27, 20.

Ferguson, R. F. 1996. Shifting challenges: Fifty years of economic change towards Black-White earnings equality. In *An American dilemma revisited: Race relations in a changing world,* edited by O. Clayton, Jr. New York: Russell Sage Foundation.

Fitzgerald, J. 1995. *Making school-to-work happen in inner cities: A white paper prepared for the MacArthur Foundation.* Chicago: Great Cities Institute, University of Illinois.

Fitzgerald, J., and D. Jenkins. 1997. *Best-practice in community colleges: Connecting the urban poor to education and employment opportunities.* Chicago: Great Cities Institute, University of Illinois.

Freeman, R. B., and L. B. Katz. 1994. Rising wage inequality: The United States vs. other advanced countries. In *Working under different rules,* edited by R. B. Freeman. New York: Russell Sage Foundation.

Gordon, D. M. 1971. *Theories of poverty and underemployment.* Lexington, MA: D. C. Heath.

Gordon, D. M., R. Edwards, and M. Reich. 1982. *Segmented work, divided workers.* New York: Oxford University Press.

Gottschalk, P., and R. Moffitt. 1994. The growth of earnings instability in the U.S. labor market. *Brookings Papers on Economic Activity* 2:217-272.

Harrison, B. 1972. *Education, training and the urban ghetto.* Baltimore: Johns Hopkins University Press.

———. 1994. *Lean and mean: The changing landscape of corporate power in the age of flexibility.* New York: Basic Books.

Harrison, B., M. Weiss, and J. Gant. 1995. *Building bridges: CDCs and the world of employment training.* New York: Ford Foundation.

Henkoff, R. 1993. Companies that train best. *Fortune,* March 22, 62.

Hershey, R. D., Jr. 1995. Working earnings post rise of 2.7%, lowest on record. *New York Times,* November 1, A1.

Holzer, H. J. 1996. *What employers want: Job prospects for less-educated workers.* New York: Russell Sage Foundation.

Howell, D. 1994. The skills myth. *American Prospect* (Summer):81-89.

———. 1997. The collapse of low-skill wages: Technological shift or institutional failure? Public Policy Brief, Jerome Levy Economics Institute, Bard College, February.

Jacoby, S. M. 1985. *Employing bureaucracy: Managers, unions, and the transformation of work in American industry, 1900-1945.* New York: Columbia University Press.

Kirschenman, J., and K. M. Neckerman. 1991. "We'd like to hire them, but . . . ": The meaning of race to employers. In *The urban underclass,* edited by C. Jencks and P. E. Peterson. Washington, DC: Brookings Institution.

Knocke, D., and A. L. Kalleberg. 1994. Job training in U.S. organizations. *American Sociological Review* 59:537-546.

Krugman, P. 1992. The right, the rich, and the facts. *American Prospect* (Fall):19-31.

Levy, F. 1995. Income and income inequality since 1970. In *State of the union: America in the 1990s.* Vol. 1, *Economic trends,* edited by R. Farley. New York: Russell Sage Foundation.

Levy, F., and R. J. Murnane. 1992. Where will all the smart kids work? *Journal of the American Planning Association* 58(Summer):283-287.

Lynch, L. M. 1993. Payoffs to alternative training strategies at work. In *Working under different rules,* edited by R. B. Freeman. New York: Russell Sage Foundation.

Mangum, G. 1995. Nostra culpae: A critique of 33 years of employment, training, and welfare programs. In *The harassed staffer's guide to employment and training policy,* edited by M. Pines, G. Mangum, and B. Spring. Baltimore: Sar Levitan Center for Social Policy Studies, Johns Hopkins University.

Mangum, S. 1995. The employment outlook for troubled populations. In *The harassed staffer's guide to employment and training policy,* edited by M. Pines, G. Mangum, and B. Spring. Baltimore: Sar Levitan Center for Social Policy Studies, Johns Hopkins University.

Marcotte, D. 1994. *Evidence of a fall in the wage premium of job security.* De Kalb: Center for Governmental Studies, Northern Illinois University.

Marshall, R., and M. Tucker. 1992. *Thinking for a living: Education and the wealth of nations.* New York: Basic Books.

Mishel, L. 1995. Rising tides, sinking wages. *American Prospect* (Fall): 60-64.

Mishel, L., and J. Bernstein. 1994. *The state of working America, 1994-95.* Armonk, NY: M. E. Sharpe.

Mishel, L., and R. A. Teixera. 1991. The myth of the coming labor shortage. *American Prospect* (Fall):98-103.

Morales, R., and F. Bonilla, eds. 1993. *Latinos in a changing U.S. economy.* Newbury Park, CA: Sage.

Moss, P., and C. Tilly. 1996a. Informal hiring practices, racial exclusion, and public policy. Unpublished manuscript, Department of Policy and Planning, University of Massachusetts at Lowell, November.

———. 1996b. "Soft" skills and race: An investigation of Black men's employment problems. *Work and Occupations* 23:252-276.

Osterman, P., ed. 1984. *Internal labor markets.* New York: Oxford University Press.

———. 1988. *Employment futures: Reorganization, dislocation, and public policy.* New York: Oxford University Press.

———. 1994. How common is workplace transformation and how can we explain who does it? *Industrial and Labor Relations Review* 47(January):173-188.

———. 1995. Skills, training, and work organization in American establishments. *Industrial Relations* (April):125-146.

Osterman, P., and R. Batt. 1993. Employer centered training programs for international competitiveness: Lessons from state programs. *Journal of Policy Analysis and Management* 12(Summer):456-477.

Owens, M. L. 1996. *Race, place, and government employment.* Albany: Nelson A. Rockefeller Institute of Government, State University of New York.

Peck, J. 1996. *Work place: The social regulation of labor markets.* New York: Guilford.

Pfeffer, J., and J. N. Baron. 1988. Taking the workers back out: Recent trends in the structuring of employment. In *Research in organizational behavior,* Vol. 10, edited by B. M. Staw and L. L. Cummings. Greenwich, CT: JAI.

Pines, M., et al. 1995. *The harassed staffer's reality check.* Baltimore: Sar Levitan Center for Social Policy Studies, Johns Hopkins University.

Piore, M. J., and C. F. Sabel. 1984. *The second industrial divide.* New York: Basic Books.

Pouncy, H., and R. B. Mincy. 1995. Out of welfare: Strategies for welfare-bound youth. In *The work alternative: Welfare reform and the realities of the job market,* edited by D. S. Nightingale and R. H. Haverman. Washington, DC: Urban Institute Press.

Rose, S. 1995. *The decline of employment stability in the 1980s.* Washington, DC: National Commission on Employment Policy.

Rosenfeld, S. A., ed. 1995. *New technologies and new skills: Two-year colleges at the vanguard of modernization.* Chapel Hill, NC: Regional Technology Strategies.

Rosenfeld, S. A., and M. E. Kingslow. 1995. *Advancing opportunity in advanced manufacturing: The potential of predominantly two-year colleges.* Chapel Hill, NC: Regional Technology Strategies.

Rosenthal, N. H. 1995. The nature of occupational employment growth: 1983-93. *Monthly Labor Review* (June):45-54.

Sassen, S. 1989. America's immigration problem. *World Policy* 6:811-832.

Savner, S. 1996. *Devolution, workforce development, and welfare reform.* Washington, DC: Center for Law and Social Policy.

Thompson, J. P. 1997. *Universalism and deconcentration: Why race still matters in poverty and economic development.* New York: Department of Political Science, Barnard College.

U.S. Bureau of the Census. 1992. *Money income of households, families, and persons in the United States: 1991.* Current Population Reports, Consumer Income, Series P-60, No. 180. Washington, DC: Government Printing Office.

———. 1993. *1990 Census of Population: Characteristics of the Black population.* CP-3-6. Washington, DC: Government Printing Office.

U.S. Department of Labor. 1995. *What's working (and what's not): A summary of research on the economic impacts of employment and training programs.* Washington, DC: U.S. Department of Labor.

Waldinger, R. 1996. *Still the promised city? African-Americans and new immigrants to postindustrial New York.* Cambridge, MA: Harvard University Press.

Waquant, L. J. D. 1994. The new urban color line: The state and fate of the ghetto in post-Fordist America. In *Social theory and the politics of identity,* edited by C. Calhoun. Oxford: Basil Blackwell, 1994.

Wolff, E. N. 1995. How the pie is sliced: America's growing concentration of wealth. *American Prospect* (Summer):58-64.

Zemsky, R. 1994. *What employers want: Employer perspectives on youth, the youth labor market, and prospects for a national system of youth apprenticeships.* Working Paper WP22. Philadelphia: National Center on the Educational Quality of the Workforce, University of Pennsylvania.

3

Why Local Economic Development and Employment Training Fail for Low-Income Communities

Elizabeth J. Mueller
Alex Schwartz

Mainstream practice in the field of local economic development and workforce development has yielded disappointing results for residents of low-income communities. In this chapter we outline dominant strategies currently in use and review evidence regarding their effectiveness in increasing employment of low-income residents. On the whole, we find that current practice, and the bulk of available public resources, is directed toward strategies that fail for residents of low-income communities. From the evaluation literature, we identify several reasons for this failure.

First, strategies typically focus on one aspect of the employment problem, under the assumption that this is the key to setting other changes in motion. Whether it be physical proximity to jobs, job search assistance, or tax costs, such a narrow focus, reflected also in inflexible program rules, precludes linkages among agencies concerned with other aspects of the employment issue and fruitful collaborations across agencies and stakeholders. It is also inconsistent with what we know about the barriers to employment of disadvantaged persons.

Second, the goals of various strategies are not consistent: Some programs aim solely at access to jobs, whereas others are concerned with the quality of jobs or the particular group of people who get them. This lack of consistency in goals, an indication of lack of political consensus around the purpose of economic development initiatives, also precludes better coordination of existing programs. In the end, success is too often defined simply as employment, even when the jobs obtained may not pay enough to move workers out of poverty.

In this chapter we first describe strategies and tools aimed at encouraging employers to maintain, expand, or create new employment in cities. These incentives typically focus on lowering specific factor costs for employers, most often land costs through abatements of local taxes. In some cases, incentives are targeted at employment of disadvantaged residents or investment in low-income neighborhoods. Based on available evidence, we find that such strategies have not produced many benefits for low-income residents. We interpret the lack of accounting and evaluation of net benefits of such incentives to be evidence of the weak commitment of city agencies to employment as a key goal. Next, we examine strategies aimed at giving residents access to jobs. These strategies have centered primarily on the provision of employment training, that is, training in job search and comportment skills rather than occupation-specific training. More recently, emphasis has been placed on strategies aimed at giving residents physical access to jobs by moving them into suburban areas or providing transportation to such areas. These programs are still small in scale and have not yet been evaluated adequately. The preliminary evidence is inconclusive; it is not yet possible to conclude whether access to suburban jobs will yield the higher wages hoped for or leave workers no better off after commuting costs than if they had taken low-wage center-city jobs.

Strategies Aimed at Employers: Development Incentives

As public resources for direct employment strategies have evaporated, strategies focused on private sector development have become the centerpiece of local economic development practice.[1] Strategies for creating jobs through private sector economic development range from development or tax incentives aimed at lowering land, capital, or labor costs for firms to more strategic regional efforts aimed at retaining employment or fostering growth in particular sectors of the regional economy through less tangible incentives (Eisinger 1988; Sternberg 1991). Tax abatements are among the most common tools used to lower land costs for investors, as are industrial revenue bonds for capital costs and wage subsidies of various sorts for labor costs. Federal strategies such as Urban Development Action Grants (UDAGs), Community Development Block Grants (CDBGs), and enterprise and empowerment zones have incorporated cost incentives directly or have provided flexible funding allowing for their use. State and local governments also frequently offer nonfinancial assistance to businesses aimed at providing information or skills to investors (Bartsch 1989). In strategic regional efforts, a recent addition to practice in regions strongly affected by secular shifts in employment, local governments act as conveners, bringing together representatives of business, labor, and government to foster more collaborative relationships and coordinated use of public resources. The pervasive philosophy across these examples is that government should play a supportive role in fostering sustained private sector activity to generate employment for city residents. Strategies have centered on "public-private partnerships" in which private sector actors have a strong hand in shaping the development process and outcomes (Barnekov and Rich 1989; Squires 1989).

Traditional economic development incentives, such as tax abatements and subsidies to investors, have been questioned by critics on several grounds (Barnekov and Rich 1989). First, such policies tend to remove economic development from the public agenda by creating quasi-public institutions to manage development and by giving private developers a significant role in these new institutions. Shielded from public scrutiny, projects (or "deals") are more likely to reflect a private sector agenda; private projects are leveraged with public dollars yet produce relatively little social benefit (Fainstein 1994). Rarely are officials or developers asked to provide evidence that the net benefits of projects, in terms of jobs or revenues, will be positive or will flow to neighborhood or city residents (Squires 1989).

Second, these traditional supply-side subsidies are criticized on equity grounds. Often, they are funded by methods that shift project costs from companies to residents while benefits flow to companies and, often, across city lines to the suburbs. With development deals, city residents come to bear the risks associated with projects and the often high costs of debt service for infrastructure and services associated with development (Lynch, Fishgold, and Blackwood 1996; Squires 1989). With revenue bonds and tax breaks, the costs and benefits of development and their jurisdictional incidence are often obscure.

Most important, neither tax abatement nor industrial development strategies are generally aimed at recruiting disadvantaged workers for jobs. They work instead on the theory that growth in employment will trickle down to such workers. Efforts to link job training and placement efforts to economic development activities are more the exception than the rule (Fitzgerald 1993). Instead, cities have focused either on office, retail, and other real estate development in the central business districts or, more recently, on recruitment of high-tech firms, with little assessment of the match between these sectors and the cities' local economy or labor market. The latter trend has persisted in spite of warnings against the replicability of high-tech growth models (Saxenian 1989) and the saturation levels reached by initiatives such as research parks and technology transfer incentives (Lugar and Goldstein 1990; Peddle 1993).

These criticisms are supported by the small but growing evaluative literature. Evaluations have cast doubt on the efficacy of supportive local economic development policies as generators of tax revenue for city coffers or jobs for city residents (Wolman and Spitzley 1996; Furdell 1994; Judd and Parkinson 1990). However, the evaluative literature is itself problematic. Most studies have suffered from serious methodological problems, often rooted in a lack of reliable data, making it difficult to draw conclusions about job creation or to attribute benefits to particular policies or projects (Robinson and Wren 1987, cited in Foley 1992). Studies measuring fiscal impacts are rare, owing in part to a general lack of rigorous accounting by local officials of abatements and purported results (Krumholz 1991; Long 1986; Giloth 1992). Attributing plant siting decisions or employment growth to particular initiatives is always extremely difficult, given the number of confounding factors and the difficulty of identifying or constructing appropriate control groups to establish what might happen in the absence of such intervention. Studies measuring employment effects too often evaluate data at the state (or metropolitan) level, yielding no information about whether city residents get new jobs, presumably the rationale for use of city-level funds in the first place.[2] In addition, practitioners and evaluators employ a range of definitions and measures of successful job creation/employment, making it hard to compare the effectiveness of various strategies (Giloth 1992; Center for Community Change 1990). Finally, the evidence available for local industrial development strategies—mainly from case studies—generally focuses on issues such as regional competitive advantage in key sectors or entrepreneurship and small business growth; employment of low-income residents in particular jurisdictions has not been seen as a central concern (Mt. Auburn Associates 1995).[3]

The evidence available on the outcomes of local economic development strategies popular in the late 1980s is discouraging. Evaluations show that most UDAG money went into downtown hotels, office buildings, and trendy festival markets (Jacobs and Roistacher 1980), creating low-wage, often part-time jobs with few benefits (Krumholz 1991). Rather than helping attract investment, at times, "UDAG grants actually followed a developer's announced intention to build" (Krumholz 1991). Numerous studies have noted that evidence on the effectiveness of tax abatements is illusory: In spite of claims to positive fiscal impacts, few states or cities have even kept records on tax abatements granted (Center for Community Change 1990; Lynch, Fishgold, and Blackwood 1996; Marlin 1990). The public is certainly not informed of the balance (Krumholz 1991).

Using the limited information available on outcomes, Lynch, Fishgold, and Blackwood (1996) conducted a study of New York State's Industrial Development Agency. They found that the agency's abatements and bonding activities caused a net loss of $1.3 billion in tax revenues between 1987 and 1991, and that although the costs of the program were readily identified, the benefits were "questionable." Attributing any increase in employment to the program was difficult: Only 2 of 23 firms moving to the state cited tax benefits as a factor in their decisions, and both stated that "they may have invested in New York State even in the absence of IDA benefits" (62). Similarly, in reviewing local economic policy and job creation in Britain and the United States, Foley (1992) found that many projects include such "deadweight costs"—that is, investment of public money that the private sector would have invested otherwise. Debt or tax increment financed deals have been further charged with shifting risks for investments to the public (Krumholz 1991).

Targeted Development Incentives

Programs aimed at generating general job or revenue growth show little effect on employment, let alone employment of low-income residents, but what about more targeted efforts? Two basic approaches to targeting have been tried: targeting benefits to particular groups of disadvantaged workers and targeting benefits to low-income neighborhoods of a city.

The Targeted Jobs Tax Credit (TJTC) was the most recent large-scale attempt to make the connection between employer incentives and the employment of low-income city residents. The credit was created in 1977 to encourage private businesses to hire disadvantaged job seekers, although it did not require employers to pay more than minimum wage. Employers received a tax credit for hiring individuals from one of several groups: disadvantaged youths, Vietnam-era veterans, ex-offenders, recipients of welfare, vocational rehabilitation clients, and high school cooperative extension students (Lorenz 1995). Businesses were initially entitled to a maximum tax credit of $4,500—50% of the first $6,000 earned in the first year and 25% of the first $6,000 earned in the second year. The Tax Reform Act of 1986 reduced the credit to $2,400—40% of the first $6,000 earned in the first year of employment. The program ended in 1994.

The designers of the TJTC program hoped that firms would be more inclined to hire disadvantaged workers if the government subsidized some of the costs of their employment. The program, however, failed to generate significant amounts of new employment. It never achieved high levels of participation, and most of the companies that did participate claimed the credit for employees they had already hired (Bishop and Kang 1991; Bishop and Montgomery 1993; Lorenz 1995). According to Bishop and Kang, the "TJTC is perhaps the most outstanding example of an entitlement program with extremely low participation rates despite a very generous offer" (1991, 25).

The evaluation literature identifies several flaws in the TJTC program's design and implementation that contributed to its low level of usage and meager job generation. First, it was difficult for businesses to recruit eligible workers actively because it is unlawful to ask job applicants whether or not they are "disadvantaged" (Bishop, 1995). The only way to do so would be for businesses to request referrals from designated state employment offices. Second, it was stigmatizing for disadvantaged workers to approach private businesses directly. Although these job seekers could be certified as eligible for the TJTC by state employment offices, such certification often undermined the job searches of these individuals because it signified to employers that they were problematic in some way.

To increase participation rates, private consultants emerged to help companies maximize their use of the credit by identifying which of their recent hires were eligible and expediting their certification by government authorities (Lorenz 1995). As a result of this practice, the vast majority of the workers claimed for the TJTC would have been hired anyway. According to Bishop and Montgomery (1993), fewer than 3 in 10 jobs benefiting from the program were new.

Besides its low participation rates and poor record of job generation, it is also important to examine the TJTC program's capacity to augment low wages beyond the poverty line. Unfortunately, the program did not require employers to offer any particular salary. They could, and often did, hire at the minimum wage. In such cases, the effective cost of hiring eligible workers was below the minimum wage. To the extent that the minimum wage discouraged employers from hiring young adults and other disadvantaged workers, the TJTC created employment opportunities for some job seekers. Even so, the program did little to boost earnings beyond the poverty line (although the minimum wage was significantly higher in real dollars at the start of the program than it is today). More important, the TJTC was temporary. Originally, employers could receive the tax credit for only 2 years per eligible worker, and after 1986 this was reduced to 1 year.[4]

A range of programs have tried to channel benefits to low-income residents by targeting subsidies to businesses operating in distressed city neighborhoods, by employing residents of such neighborhoods, or both. The assumption is that most new jobs generated in low-income neighborhoods will go to neighborhood residents. In some cases, benefits are tied to local hiring.

The most prominent recent example of geographic targeting is the group of state programs aimed at particular low-income or blighted neighborhoods that are designated as *enterprise zones*. Although particular program models vary by state, most states establish enterprise zones in areas that are determined to be "economically distressed," and offer special incentives and benefits to the targeted areas in order to spur new investment and job creation (Dabney 1991). Common incentives include reduced property taxes, reduced utility costs, and local sales tax reductions (Elling and Sheldon 1991).[5] Over time, state enterprise zones have broadened their geographic focus and have become located increasingly in suburban or rural areas. They more often focus on job retention in the context of shifting regional economies, in contrast to the initial focus on job creation (Wolf 1990). The federal Empowerment Zone and Enterprise Communities Initiative builds on the enterprise zone concept, offering place-based incentives in the context of a local planning process, with community oversight of the allocation of economic development incentives.

Current evaluations of empowerment zones have not yet yielded evidence on their employment effects, but a large critical literature has followed the unfolding of state enterprise zone programs, documenting the programs' mixed results. The heart of the debate surrounding enterprise zones, again, is the importance of tax costs in business location decisions. From his analysis of Dun & Bradstreet data for business activity in zones in eight states, Dabney (1991) concludes that tax incentives are offset by other costs. On the whole, he notes that enterprise zone incentives have marginal effects on business location decisions except in those cases where the value of the incentive is large relative to the amount of the investment. For this reason, small businesses are most likely to be attracted by such initiatives. According to Dabney, "The typical enterprise zone program consisting of limited tax incentives and regulatory relief is unable by itself to adequately address the necessary factors found to be important in most business location decisions" (1991, 334).

In a similar vein, Dowall (1996) studied the influence of California's enterprise zone program on plant location and found little effect. Under the two state enabling laws, zone businesses are eligible for tax incentives for the purchase of manufacturing equipment and for hiring employees. Localities typically have added local incentives such as job training or financing assistance. In some cases, loan funds have been set up. Funds have been derived from CDBG or state redevelopment funds or other state or local general-obligation or revenue bonds supporting public infrastructure or private development. Like Dabney, Dowall (1996, 364) found that the net effect of various zone credits was modest and unable to overcome the many other factors influencing business investment and location decisions. Only 23% of surveyed firms reporting use of enterprise zone incentives indicated that such incentives had influenced their plant location or expansion decisions.

Evidence on the net employment and earnings effects of enterprise zones is also discouraging. By 1995, 40 studies of such effects had been conducted, using a wide range of methodologies and differing outcome measures, and yielding inconclusive results (Dowall 1996). The strongest results have been produced by studies that rely on information gathered from businesses or zone administrators on the impacts of zone activities on employment. In effect, survey respondents claim that they do respond to tax incentives (Dowall 1996), yet less than half of zone businesses report using such incentives. Sridhar (1996) has concluded that zones are beneficial by calculating the benefits of zone employment based on the increase that wages offer over estimated reservation wages—a method that may exaggerate gains.

More qualitative researchers have pointed out the significance of zone management and other nonquantifiable factors in the success of zones (Alley et al. 1993). They also emphasize the importance of matching tax incentives with investments in human and physical infrastructure, including transportation services and facilities, police protection, and education and training systems (Levitan and Miller 1992). Hornbeck (1993) makes some specific recommendations that are consistent with those of several other researchers. He proposes that zones be designed so that benefits are closely targeted to intended beneficiaries following five guidelines:

1. Select areas meeting a minimum threshold of resources necessary for possible growth.
2. Design financial incentives to have significant effects on firm costs in the short run.
3. Require that zones compete for benefits.
4. Target incentives toward firms that are likely to create long-term job prospects.
5. Require zones to have coordinated management.

All the evidence is not yet in on the employment effects of federal empowerment zones, but we can surmise something about their prospects from the

BACKGROUND

economic development tools they provide. The federal empowerment zone program offers an Employer Wage Credit that is similar in some respects to the TJTC. Eligible private businesses operating in the six federally designated empowerment zones can claim a tax credit of 20% for the first $15,000 of a qualified worker's wages. After 2001, the credit diminishes by 5% each year until 2004 (U.S. Department of Agriculture and U.S. Department of Housing and Urban Development 1994, 69-70). The empowerment zone employer wage subsidy is available for a longer period than the TJTC, which could never be claimed for more than 2 years for any eligible worker. The Employer Wage Credit also differs from the TJTC in that it limits the geographic location of eligible employers—establishments claiming the credit must be situated within designated empowerment zones. Otherwise, the credit is similar in concept to TJTC. Although the empowerment zone subsidy does not explicitly target "disadvantaged" workers, as the TJTC did, the credit can be claimed only for employees who are zone residents, and therefore are presumably "disadvantaged." Most important, the Employer Wage Credit, like the TJTC, does not require employers to pay above the minimum wage. Neither wage subsidy program ensures that workers' earnings exceed the poverty line.

In sum, the most commonly used tools of economic development practice are incentives focused on factor costs that may not be of central importance to businesses in their location and hiring decisions. In any case, they have not been effectively targeted at low-income residents and, as a result, have likely yielded few employment benefits for them.

Strategies Aimed at Workers: Access and Skills

Government agencies and nonprofit organizations have used several strategies to help inner-city residents gain access to employment. Most often, these strategies have involved employment training and other educational programs. More recent strategies have used housing and transportation to make skill-appropriate jobs more accessible to inner-city residents. In this section we review the experience to date with these strategies.

Employment Training

Currently, publicly funded job training is based on the provision of short-term training, primarily in job seeking or in a nexus of skills labeled *job readiness*.

Any occupational training included is of extremely brief duration. This approach is based on the premise that unemployment is due to cyclical or frictional factors rather than structural barriers driven by labor market changes or discrimination. Many analysts have challenged this premise, pointing out that the current labor market does not provide enough jobs that pay wages high enough to enable trainees to support themselves or make the transition off of public assistance in the long run (Lafer 1991; Edin 1994). The Clinton administration's "one-stop shop" approach to reforming the training system only serves to rationalize it and does not fundamentally alter its approach to unemployment. Yet evidence from evaluations of current and past job training programs gives us little reason to be optimistic about job training as an effective antipoverty strategy.

The United States has never funded job training (using the broadest of definitions) at a level comparable to other advanced industrialized countries. Even the Comprehensive Employment Training Act (CETA) program of the 1970s, which was funded at 10 times the level of the current federal job training program, never funded job training on a large scale. In fact, most of CETA's budget, $11 billion annually at its peak, was directed at creating jobs through public sector employment, on the theory that the private sector could not generate enough jobs in the near term (Lafer 1991, 369). With the election of Ronald Reagan, this approach was abandoned. Unemployment, especially among the poor, was redefined as a supply-side issue, attributed to the characteristics of the unemployed rather than those of the market.

The Job Training Partnership Act (JTPA), which replaced CETA after its authorizing legislation expired in 1982, embodies a completely different orientation to the issue of jobs for the poor. The JTPA system treats the unemployment of middle- and low-income people differently, attributing the former to cyclical downturns and the latter to lack of skills or job readiness. The solutions offered for cyclical unemployment are job search assistance and unemployment insurance; for the longer-term unemployment of the poor, the solutions are job readiness and short-term skills training. The program carefully avoids any public sector job creation.

The scale of JTPA is modest. In 1993, JTPA Title II, which serves disadvantaged workers, enrolled 310,000 adults (plus 140,000 youth) at a cost of $1.7 billion (U.S. Department of Labor 1995, 28). Most commonly used services under the program include classroom training (46%), job search assistance (19%), and on-the-job training (18%). Classroom training can include either occupational or basic skills training (U.S. Department of Labor 1995).

Criticisms of the program are based on many features, ranging from its underlying conception to its implementation. The program's reliance on dated labor market statistics in the current context of rapid change can lead to inappropriate training. Practitioners often complain that the system is poorly linked to local employers and that training developed to lead to previously identified jobs at local companies is de-emphasized because of the danger of subsidizing the training that employers would have provided in any case (Oden et al. 1993). In addition, such training is usually aimed at "displaced" rather than "disadvantaged" workers. Due to labor market conditions and the program's emphasis on linking payment to service providers to performance indicators (such as placement rates), the program pushes local contractors to choose the job ready over the needy for inclusion in JTPA programs (Donahue 1989). In response, program design was altered in 1992 to increase targeting of low-income people who face multiple barriers to work. Unfortunately, systematic evaluations based on post-1992 program data are not yet available (U.S. Department of Labor 1995, 30).

Evidence on the effectiveness of JTPA's program for "disadvantaged" workers (Title II-A) is not encouraging. A recent review of evaluations of job training programs conducted by the Department of Labor paints a discouraging picture. Overall, although higher earnings have been achieved for adult women and men, they have been due mostly to increased hours of work rather than higher hourly wages. In fact, participants' annual earnings still place them well under the poverty threshold. Not surprisingly, welfare receipt has not decreased as a result of participation. The program elements found to be most effective in producing these increases are job search and on-the-job training. Women have tended to show larger increases in annual income than have men, and welfare recipients have shown the greatest increases (although not great enough to move them off of welfare) (U.S. Department of Labor 1995, 27-32).

In response to the poor results of Title II programs and calls for welfare reforms emphasizing the transition to work, experiments with different types of programs were implemented in the 1980s and early 1990s. These include federally funded demonstration projects and state-run welfare reform projects, the latter made possible by federal waivers of some welfare regulations.

There have been three national demonstrations: the Minority Female Single Parent Demonstration (MFSP), the Supported Work Demonstration (SWD), and the Home Health Aide Demonstration (HHA). The MFSP offered job skills training, basic education, and supportive services to approximately 4,000 single parents at four sites.[6] Two-thirds of those enrolled were welfare recipients. The SWD and HHA each provided about a year of subsidized employment to single parents who were long-term welfare recipients. The SWD was an expensive, intensive program; the HHA was less expensive and provided fewer services to participants. Both programs were found to produce gains in annual income of about $1,700 in the first 2 years after participation, gains beyond those achieved by control group members. The HHA produced the largest gains. Yet again, the results had little effect on poverty, as participants' incomes remained quite low (U.S. Department of Labor 1995).

Only one site in the MFSP demonstration produced positive results—the Center for Employment Training (CET) in San Jose, California. This program has become a model for a series of replication projects across the country, some of which are run by community-based organizations. The CET model involves integration of basic skills training into a vocational skills curriculum. Although the program has been shown to raise earnings substantially again, posttraining earnings have remained below the poverty level ($7,080 per annum 2 years after program end) (Mathematica Policy Research 1993).

■ State-Level Welfare Reform and Training

Several states have developed welfare programs that feature job training services as a key component. Under the Work Incentive (WIN) program in the 1980s, some states provided job search and unpaid work experience. The Family Support Act of 1988 expanded these efforts under the Job Opportunities in the Business Sector (JOBS) program, which provided basic education and training along with extensive job search assistance. In contrast to the demonstrations described above, these programs were mandatory—participants were not self-selected volunteers. Overall, these programs have reported modest gains in income, again usually due to increased hours of work rather than increased hourly wages. In addition, program participation appears to have had little if any impact on the poverty, unemployment, or welfare receipt rates of participants (Riccio et al. 1994; U.S. Department of Labor 1995, 36-37). Longer-term evaluations of these programs, conducted over 5 years, highlight the importance to sustained earnings gains and self-sufficiency of placing participants in higher-wage jobs. Without such jobs, the gains of programs have been found to fade over time (Friedlander and Burtless 1995).

Programs emphasizing more general educational attainment (rather than occupational or job readiness training) show minimal results until the college level. Beyond this threshold, there is a significant earnings payoff. Below this level, programs aimed at basic education and/or GED show marginal results (U.S. Department of Labor 1995).

■ Training for Better Jobs

Due to the inability of federally funded programs to generate workers' access to better-paying jobs or jobs with the potential to lead to better wages in the future, community-based organizations and foundations active in the area of antipoverty work have been searching for alternative approaches. These organizations have established programs that are aimed at providing more intensive skills training, better postplacement services, and, in the most ambitious programs, jobs that offer above-poverty wages or that put workers on a career track. Given the paucity of such jobs in inner cities (and elsewhere), this often means creating new jobs or extended on-the-job training, in promising occupations, within a business run by the community-based training organization itself. Such approaches are not easily funded under federal programs and require substantial start-up capital and extensive training periods to achieve their goals (McGill 1996).

Overall, the evidence on the results of federally funded job training programs for inner-city workers is not encouraging. Increases in income do not raise participants above the poverty threshold. Most evaluations have found that any increases in income are most likely to come from increased work effort rather than from increased wages. This suggests that workers remain in the same extremely low-wage jobs as before, only working longer hours. Alternative programs developed by nonprofit organizations with foundation support are in the early stages, and have not yet been adequately evaluated. To the extent that they are able to develop jobs in higher-wage occupations that require relatively low levels of skill, they present a more viable model.

Housing and Transportation-Based Strategies

Two related strategies attempt to link city residents with suburban employment opportunities: dispersal and mobility. Residential dispersal strategies try to help city residents overcome the institutional and economic barriers that deny them access to suburban housing and the opportunities for education and employment such housing offers. Mobility strategies also try to connect city residents with suburban jobs, but without requiring residents to move, thus sidestepping the protracted and bitter opposition that often forms in reaction to efforts to "open the suburbs" to low-income and minority families.

■ Residential Dispersal

Residential dispersal strategies increase the accessibility of suburban areas to city residents through such means as enforcement of fair housing laws, inclusionary zoning, scattered-site public housing, elimination of regulatory barriers to affordable housing, and broadening of the residential choices for holders of rental vouchers (i.e., tenant-based Section 8 housing subsidies). As an economic development strategy for central-city residents, a dispersal plan assumes that residential proximity to jobs will help formerly unemployed city residents secure employment.

The dispersal strategy is best exemplified by Chicago's Gautreaux Assisted Housing Program, a court-ordered remedy to the racial policies of the Chicago Housing Authority and the U.S. Department of Housing and Urban Development (HUD). The Gautreaux program offers Section 8 rental certificates along with counseling, placement, and other support for public housing residents and families on public housing waiting lists to relocate to private housing in selected middle-income neighborhoods in the city of Chicago and in selected suburbs. Participants are selected by lottery and subsequent screening. Since 1975, more than 5,000 families have participated in the program, more than half of whom moved to middle-income, predominantly white suburbs (Rosenbaum 1995; Davis 1993).

The strategy is also embodied in HUD's $234 million Moving to Opportunity (MTO) demonstration program. Started in 1994, MTO provides housing vouchers and other types of assistance to public housing residents in five cities: Baltimore, Boston, Chicago, Los Angeles, and New York. MTO assigns participants into three groups. Members of the "experimental group" are required to move to census tracts with poverty rates of less than 10%. They receive Section 8 housing certificates along with intensive counseling services from a nonprofit organization, including landlord recruitment, guidance in conducting housing searches, and home visits. The second group also receives Section 8 subsidies but does not get any special counseling. The third group does not receive Section 8 subsidies or counseling and is expected to remain in public housing (Polikoff 1994).

Preliminary findings are not yet available for the MTO demonstration project. Research on Chicago's Gautreaux program, however, offers some evidence to suggest that dispersal increases employment opportunities for city residents. Rosenbaum's studies of Gautreaux indicate that participants who move to the suburbs are more likely to find jobs than are participants who move within the city limits. For example, 46% of the suburban movers who were unemployed before the move found employment, compared with just 30% of the city movers who were unemployed before moving. However, Rosenbaum (1995; Rosenbaum and Popkin 1991) found no difference between suburban and city movers in average wages or hours worked. In other words, moving to the suburbs did not, by itself, increase participants' financial benefits from work. It should also be pointed out that the Gautreaux program is small in scale and extremely labor-intensive to administer. Only a few hundred residents participate each year, and each receives extensive counseling. Studies of MTO should show how important housing counseling, landlord recruitment, and other assistance is to the program's success in relocating public housing residents to the suburbs. If the program were much larger, administrators would be hard-pressed to provide the same degree of counseling, and they would probably face stiffer local opposition to the relocation of ghetto residents (Hughes 1995; Davis 1993; Polikoff 1994).[7]

■ Mobility

The mobility strategy potentially offers a more direct way of connecting city residents with suburban jobs. Rather than relocating city residents to the suburbs and hoping they can then find jobs, the mobility strategy aims to connect city residents with suburban jobs through transportation improvements and other support services. HUD is testing the mobility strategy in its Bridges to Work demonstration project, a 4-year, $18 million effort to link inner-city welfare recipients in five cities to suburban job opportunities.

Several state and local governments and nonprofit agencies have also established van service and other reverse-commuting programs to help inner-city residents access suburban jobs. For example, the South Eastern Pennsylvania Transportation Authority's Bus-Rail Link program provides bus connections from suburban rail stations to major suburban employers. Employers help to subsidize the program's operating costs. Wisconsin's Job-Ride program sponsors van service from inner-city neighborhoods to suburban job sites (Hughes 1993).

Chicago's nonprofit Suburban Job Link Corporation is probably the nation's oldest reverse-commuting program for inner-city residents. Founded in 1971, it links unemployed city residents with blue-collar manufacturing jobs in the city's suburbs. The nonprofit organization operates express bus service from Chicago's west side to job sites in the northwest suburbs. It is also developing a ride-share management program with a suburban bus company to provide vanpool service for city residents. Besides transportation, Suburban Job Link also runs a temporary employment service that places city residents in suburban jobs. In 1995, it established the Job Oasis in suburban Chicago. This facility combines work-preparedness and employment training with job referrals, placement assistance, and postplacement support. It also provides facilities where people can prepare résumés and schedule job interviews with suburban employers. Suburban Job Link provides free transportation from the inner city to the Job Oasis and from the Oasis to suburban companies for job interviews (Suburban Job Link Corporation 1996; Bardoe 1996, 19-20).

Mark Allen Hughes, the nation's leading advocate for mobility-based employment strategies and a key architect of HUD's Bridges to Work program, claims that improved transportation access to suburban employment centers would enable inner-city residents to secure jobs that offer higher pay and are better suited to their skills and backgrounds. Hughes argues that suburban business centers have much lower unemployment rates than do central cities and that suburban labor shortages translate into higher wages for entry level jobs: "Why direct people to $4/hour fast food jobs downtown if the same jobs in the suburbs pay $6/hour?" (1993, 243). Hughes also argues that suburban jobs are more appropriate for inner-city residents in terms of their skill requirements. The suburbs' metropolitan dominance in manufacturing, wholesale trade, and retailing translates into employment opportunities for people with limited education and skills.

As yet there have been few if any complete evaluations of the mobility strategy. Nevertheless, there are several empirical and theoretical reasons to question whether mobility alone will enable inner-city residents to escape poverty through employment. First, the suburban labor shortage is open to question. Hughes claims that slow employment growth projected for the 1990s and beyond will create an increasing shortage of suburban labor, thus making suburban firms more interested in hiring workers from the central city. However, heightened international migration, unparalleled since the early 20th century, has offset some of the slow growth anticipated by domestic demographic trends (Pitkin and Simmons 1996). It is also not likely that suburban employment opportunities will increase faster than the number of public assistance recipients forced into the workforce through welfare reform.

It is also debatable whether suburban entry-level jobs offer significant pay advantages over entry-level jobs located in the central city (Holzer 1994; White and McMahon 1995). Even if suburban jobs do pay somewhat higher wages, this advantage would be offset by additional commuting costs (Porter 1995, 61).

Hughes may also exaggerate the advantage suburban areas have over the central city in the suitability of entry-level jobs for low-skilled city residents. True, suburbs do have more manufacturing and retail jobs than most central cities. However, this does not mean that there is a dearth of entry-level jobs in the central city; not all, or even most, central-city jobs require advanced degrees. There are plenty of service jobs in central cities—positions such as food handler, maintenance worker, security guard, business machine operator, messenger—that require little or no education or work experience (Sassen 1993; Waldinger 1996). Indeed, several of the industries in which Hughes claims central cities have a comparative advantage, such as tourism and conventions, employ mostly low-skilled workers. The problem is less the availability of employment opportunities for inner-city residents than the availability of jobs offering decent wages and benefits (Howell 1995).

Another issue is logistical. Are reverse-commuting programs feasible in suburban areas with low employment density—where business establishments are small in scale, widely scattered, or both?

In short, mobility by itself does not ensure that city residents will obtain jobs that can lift them out of poverty. Indeed, Chicago's Suburban Job Link Corporation and the federal Bridges to Work initiative recognize this and combine transportation with other supportive services, such as training, job referrals, and postplacement support. Hughes also sees transportation as part of a broader strategy that includes an increased minimum wage and/or Earned Income Tax Credit (EITC). He notes that the EITC would "provide an important mechanism by which to lighten the burden of [transportation] costs on workers by increasing the attractiveness of low wage jobs and possibly by incorporating commuting costs in some way" (1993, 295). However, the added commuting costs inherent in the mobility strategy would reduce the ability of the EITC to increase the effective incomes of low-wage workers.

In sum, household dispersal and mobility represent relatively new strategies for improving employment among inner-city residents. They try to connect low-income city residents to jobs in the suburbs, where most of the nation's employment growth takes place. Most programs based on these strategies have begun only recently and are still being evaluated. It is too early to assess their effectiveness. Nevertheless, studies of Chicago's Gautreaux program and questions about the assumptions behind the mobility strategy raise doubts about whether these strategies, by themselves, can lead city residents out of poverty.

Conclusion

Recent practice in the field of economic and workforce development has yielded disappointing results for residents of low-income communities. Current practice, and the bulk of available public resources, is directed to strategies that fail for precisely those workers most in need of assistance. Strategies focus too exclusively on particular aspects of the employment problem, precluding more holistic approaches that are appropriate to the needs of both workers and employers. The goals of commonly pursued strategies are not consistent, with some programs emphasizing absolute increases in employment, others focusing on who gets the employment, and still others stressing the quality of jobs generated. Too few focus on whether gaining access to jobs moves disadvantaged city residents out of poverty. Finally, there is an overall lack of coordination across existing programs, reflecting the lack of consensus around program goals and the narrow organization of programs and agencies themselves. The subsequent chapters in this volume present promising emerging models for linking the needs of residents and employers.

Notes

1. Currently, there is consensus that the level of expansion that would be needed to absorb the urban unemployed is unlikely to come from the public sector. Urban workfare programs in New York and other large cities—as currently operated—are more appropriately seen as cost-saving strategies for the provision of city services than as pathways toward sustainable employment for low-income residents. Advocacy groups in New York City have charged the city with cycling recipients in and out of such jobs and with failing to provide necessary training or tenure for successful transition to private sector employment (see Community Service Society of New York, 1996). In response to such charges, the New York State Supreme Court recently handed down a decision forcing the city to stop assigning welfare recipients to workfare programs until it first does a full assessment of recipients' needs and skills to determine whether they should be put into educational programs instead ("Assess Workfare Recipients," 1997).

2. The exception is enterprise zone evaluations, which do focus on highly localized employment effects (see the discussion below).

3. Yet preliminary evidence on the work of the Garment Industry Development Corporation in New York City and the Machine Action Project and Needle Trades Action Project in Massachusetts suggests that new initiatives focused on the development of industries employing many low-skill workers can be effective in retaining or increasing employment among particular groups of low-income workers. In addition, advocates of sectoral strategies have proposed that cities set aside light manufacturing zones aimed at retention of urban manufacturers.

4. Several smaller-scale wage subsidy demonstration projects are discussed in the next section.

5. Although Congress enacted the Housing and Community Development Act of 1987, whose Title VII authorized designation of 100 enterprise zones (33 of which were to be in rural areas), the federal program languished because congressional proposals to provide development incentives have not been enacted.

6. Funded by the Rockefeller Foundation, this demonstration ran between 1982 and 1988 and was evaluated by Mathematica Policy Research.

7. Such local opposition nearly forced the federal government to terminate MTO. Residents of several blue-collar communities outside Baltimore complained vigorously to their congressional representatives and senators that the program would inundate their neighborhoods with poor residents from Baltimore's public housing projects (De Witt 1995).

References

Alley, A., R. Gittell, H. McFarland, A. Rein, A. Vidal, and M. Wilder. 1993. *The community development impacts of the Indiana enterprise zone program and the Lilly Endowment initiative on community development.* New York: Community Development Research Center.

"Assess Workfare Recipients, Judge Tells City." 1997. *New York Times,* March 26.

Bardoe, C. 1996. *Employment strategies for urban communities: How to connect low-income neighborhoods with good jobs.* Chicago: Center for Neighborhood Technology.

Barnekov, T., and D. Rich. 1989. Privatism and the limits of local economic development policy. *Urban Affairs Quarterly* 25:212-238.

Bartsch, C. 1989. Government and neighborhoods: Programs promoting community development. *Economic Development Quarterly* 3:157-168.

Bishop, J. H. 1995. Telephone interview. September.

Bishop, J. H., and S. Kang. 1991. Applying for entitlements: Employers and the targeted jobs tax credit. *Journal of Policy Analysis and Management* 10:24-45.

Bishop, J. H., and M. Montgomery. 1993. Does the targeted jobs tax credit create jobs at subsidized firms? *Industrial Relations* 33:289-306.

Center for Community Change. 1990. *Bright promises; questionable results: An examination of how well three government subsidy programs created jobs.* Washington, DC: Center for Community Change.

Community Service Society of New York. 1996. *New York City's work experience program.* Urban Agenda Issue Brief 1. (October). New York: Community Service Society of New York.

Dabney, D. Y. 1991. Do enterprise zone incentives affect business location decisions? *Economic Development Quarterly* 5:325-334.

Davis, M. 1993. The Gautreaux assisted housing program. In *Housing markets and residential mobility,* edited by G. T. Kingsley and M. A. Turner. Washington, DC: Urban Institute Press.

De Witt, K. 1995. Housing voucher test in Maryland is scuttled by a political firestorm. *New York Times,* March 28, B10.

Donahue, J. 1989. *Shortchanging the workforce: The Job Training Partnership Act and the overselling of privatized training.* Washington, DC: Economic Policy Institute.

Dowall, D. E. 1996. An evaluation of California's enterprise zone program. *Economic Development Quarterly* 10:352-368.

Edin, K. 1994. *The myths of dependence and self-sufficiency: Women, welfare, and low-wage work.* Working Paper 67. New Brunswick, NJ: Center for Urban Policy Research, Rutgers University.

Eisinger, P. K. 1988. *The rise of the entrepreneurial state: State and local economic development policy in the United States.* Madison: University of Wisconsin Press.

Elling, R. C., and A. W. Sheldon. 1991. Determinants of enterprise zone success: A four-state perspective. In *Enterprise zones,* edited by R. Green. Newbury Park, CA: Sage.

Fainstein, S. S. 1994. *The city builders: Property, politics, and planning in London and New York.* Cambridge: Basil Blackwell.

Fitzgerald, J. 1993. Labor force, education, and work. In *Theories of local economic development: Perspectives from across the disciplines,* edited by R. D. Bingham and R. Mier. Newbury Park, CA: Sage.

Foley, P. 1992. Local economic policy and job creation: A review of evaluation studies. *Urban Studies* 29:557-598.

Friedlander, D., and G. Burtless. 1995. *Five years after: The long term effects of welfare-to-work programs.* New York: Russell Sage Foundation.

Furdell, P. 1994. *Poverty and economic development: Views of city hall.* Washington, DC: National League of Cities.

Giloth, R. P. 1992. Stalking local economic development benefits: A review of evaluation issues. *Economic Development Quarterly* 6:80-90.

Holzer, H. J. 1994. Black employment problems: New evidence, old questions. *Journal of Policy Analysis and Management* 13:699-722.

Hornbeck, J. F. 1993. *Enterprise zones: Can a federal policy affect local economic development?* Washington, DC: Congressional Research Service.

Howell, D. 1995. *The collapse of low-skill wages: Technological shift or institutional failure?* New York: Robert J. Milano Graduate School of Management and Urban Policy, New School for Social Research.

Hughes, M. A. 1993. Antipoverty strategy: Where the rubber hits the road: Transporting workers to jobs. In *Housing markets and residential mobility,* edited by G. T. Kingsley and M. S. Turner. Washington, DC: Urban Institute Press.

———. 1995. A mobility strategy for improving opportunity. *Housing Policy Debate* 6:271-297.

Jacobs, S., and E. Roistacher. 1980. The urban impacts of HUD's Urban Development Action Grant program, or where's the action in Action Grants? In *Urban impacts of federal policies,* edited by N. Glickman. Baltimore: Johns Hopkins University Press.

Judd, D., and M. Parkinson. 1990. *Leadership and urban regeneration.* Newbury Park, CA: Sage.

Krumholz, N. 1991. Equity and local economic development. *Economic Development Quarterly* 5:291-300.

Lafer, G. 1991. The politics of job training: Urban poverty and the false promise of JTPA. *Politics and Society* 22:349-388.

Levitan, S. A., and E. Miller. 1992. Enterprise zones are no solution for our blighted areas. *Challenge,* May-June, 408.

Long, N. 1986. Getting cities to keep books. *Journal of Urban Affairs* 8(Spring):1-7.

Lorenz, E. C. 1995. TJTC and the promise and reality of redistributive vouchering and tax credit policy. *Journal of Policy Analysis and Management* 14:270-291.

Lugar, M., and H. Goldstein. 1990. Research parks as public investments. In *Financing economic development,* edited by R. D. Bingham, E. Hill, and S. White. Newbury Park, CA: Sage.

Lynch, R. G., G. Fishgold, and D. L. Blackwood. 1996. The effectiveness of firm-specific state tax incentives in promoting economic development: Evidence for New York State's industrial development agencies. *Economic Development Quarterly* 10:57-68.

Marlin, M. R. 1990. The effectiveness of economic development subsidies. *Economic Development Quarterly* 4:15-22.

Mathematica Policy Research, Inc. 1993. *More jobs and higher pay: How an integrated program compares with traditional programs. The Minority Female Single Parent Demonstration.* New York: Rockefeller Foundation.

McGill, D. 1996. *Community-based employment training: Four innovative strategies.* Washington, DC: National Congress for Community Economic Development and Community Development Research Center.

Mt. Auburn Associates. 1995. *Jobs and the urban poor: Publicly initiated sectoral strategies.* Vol. 2, *Case studies.* Report to the Ford and Charles Stewart Mott Foundations. Cambridge, MA: Mt. Auburn Associates.

Oden, M., J. Feldman, C. Hill, A. Markusen, and E. Mueller. 1993. *Changing the future: Converting the St. Louis economy.* Working Paper 59. New Brunswick, NJ: Center for Urban Policy Research, Rutgers University.

Peddle, M. 1993. Planned industrial and commercial developments in the U.S. *Economic Development Quarterly* 7:107-124.

Pitkin, J. R., and P. A. Simmons. 1996. The foreign born population to 2010: A prospective analysis by country of birth, age, and duration of U.S. residence. *Journal of Housing Research* 7(1):1-31.

Polikoff, A. 1994. Housing policy and urban poverty. Paper presented at the 16th Annual Conference of the Association for Public Policy Analysis and Management, October 10, Chicago.

Porter, M. 1995. The competitive advantage of the inner city. *Harvard Business Review* 73(May-June):55-71.

Riccio, J., et al. 1994. *GAIN: Benefits, costs and three-year impacts of a welfare to work program.* New York: Manpower Demonstration Research Corporation.

Robinson and Wren. 1987. Evaluating the impact and effectiveness of financial assistance policies in the Newcastle metropolitan region. *Local Government Studies* 13:49-61.

Rosenbaum, J. E. 1995. Changing the geography of opportunity by expanding residential choice: Lessons from the Gautreaux program. *Housing Policy Debate* 6(1):231-271.

Rosenbaum, J. E., and S. J. Popkin. 1991. Employment and earnings of low-income blacks who move to middle-class suburbs. In *The urban underclass,* edited by C. Jencks and P. E. Peterson. Washington, DC: Brookings Institution.

Sassen, S. 1993. *The global city: New York, London, Tokyo.* Princeton, NJ: Princeton University Press.

Saxenian, A. 1989. The Cheshire Cat's grin: Innovation, regional development and the Cambridge case. *Economy and Society* 18:448-477.

Squires, G., ed. 1989. *Unequal partnerships: The political economy of urban redevelopment in postwar America.* New Brunswick, NJ: Rutgers University Press.

Sridhar, K. S. 1996. Tax costs and employment benefits of enterprise zones. *Economic Development Quarterly* 10:69-90.

Sternberg, E. 1991. The sectoral cluster in economic development policy. *Economic Development Quarterly* 5:342-356.

Suburban Job Link Corporation. 1996. *Linking residents of inner-city neighborhoods with good employment opportunities.* Chicago: Suburban Job Link Corporation.

U.S. Department of Agriculture and U.S. Department of Housing and Urban Development. 1994. *Guidebook for community-based strategic planning for empowerment zones and enterprise communities.* Washington, DC: Government Printing Office.

U.S. Department of Labor. 1995. *What's working (and what's not): A summary of research on the economic impacts of employment and training programs.* Washington, DC: U.S. Department of Labor.

Waldinger, R. 1996. *Still the promised city? African Americans and new immigrants in postindustrial New York.* Cambridge, MA: Harvard University Press.

White, S. B., and W. F. McMahon. 1995. Why have earnings per worker stagnated? *Journal of Urban Affairs* 17(1):33-52.

Wolf, A. 1990. Enterprise zones: A decade of diversity. In *Financing economic development,* edited by R. Bingham, E. Hill, and S. White. Newbury Park, CA: Sage.

Wolman, H., and D. Spitzley. 1996. The politics of local economic development. *Economic Development Quarterly* 10:115-150.

4

Networks, Sectors, and Workforce Learning

LAURA DRESSER
JOEL ROGERS

Urban jobs systems—that is, institutionalized and publicly recognized and supported programs for preparing people for work, placing and keeping them in jobs, and providing opportunities for advancement—are notoriously weak, and almost impossible to navigate for those without insider knowledge. Skills mismatches, spatial mismatches, network-driven access systems, and inefficient training programs predominate, pushing urban job seekers away from good jobs and family-supporting incomes. Not only are these complex systems difficult to navigate from the community perspective, employers seeking good connections to potential workers also find themselves besieged by phone calls from community groups and individuals, but have no real knowledge of the quality of services offered and no single point of access to systems that could connect them to future workers. Overcoming these problems requires a systematic focus on jobs systems, on barriers to work, and on employers' needs. Only in simultaneously responding to all of these concerns can an effective urban jobs agenda be developed.

So stated, the goal is straightforward enough. Practice on the ground, however, is difficult. Given the multiple actors, agendas, needs, and responsibilities in urban jobs systems, there is no simple method of improving coordination on employer demand and community needs. Nevertheless, our experience in Wisconsin suggests that sectoral training partnerships in leading sectors can provide the foundation for real system reform. In the short term, these partnerships can develop clear routes of access into and up through industries for individuals. In the long term, as more leading sectors are organized, sectoral partnerships could collectively provide the necessary scale and detail on labor market information that effective administration of public support and training will require.

Labor Market Realities

Essentially, for labor markets to work well for people, they must provide clear and reliable methods of access to jobs and advancement through them. For example, entry-level employment should prepare workers for and connect them to future opportunities; there should be predictable, fair, and well-known methods of access to decent-paying jobs; and incremental moves and skill development on the job should allow for routine career advancement. All of this requires complete information on labor market opportunities, strong systems of training on the job, and good understanding of the relationship between the skills required for one job and those required for the next.

This rational labor market was most clearly approximated in post-World War II manufacturing. Young workers could "go down to the factory and sign up" with reasonable expectation of increasing their skills, wages, and seniority over the years. Entry-level workers took low-skill positions and banked on the promise of an orderly and secure progression to higher skill levels. Because job stability grew with seniority, incumbent workers benefited as they moved up the ladder.

Not anymore. Work restructuring, economic shifts toward service, declining firm size, and changing governance mechanisms have all hastened the collapse of this orderly system. Although internal labor markets certainly continue to exist, present work organization is widely characterized by a much less clear system. Firms have reduced the total number of job descriptions (stripping the rungs from ladders) and have cross-functionally defined job descriptions (ladders have crashed into one another). Accordingly, jobs on average carry somewhat more demanding human capital requirements, and movement across them is more driven by worker demonstration of specific skills—albeit generally skills still specific to individual firms. Additionally, as firms become more narrowly focused on "core competencies," they outsource many of the entry-level jobs that once provided routes toward the core positions. In many instances, work reorganization demands more abundant and more general skills (e.g., teamwork and flexible production) even as it eliminates the sorts of intermediate positions where those skills could be incrementally gained.

At the same time work has been reorganized, production in the economy has also been shifting: The service sector is growing and firm size is falling. These factors also work against the development of career ladders. Service sector workers never really enjoyed the orderly advancement provided in the manufacturing sector. In part, this is due to the extreme bifurcation of the service sector, which provides extremely high-paying, high-skill jobs at one end and very low-wage, low-skill work at the other. Intermediate positions and pathways out of the low-wage work have never been well developed. For example, the distance in training and pay between certified nursing assistant and licensed practical nurse is quite large, and making the transition is neither common nor convenient. Advancement in clerical careers generally requires movement from one industry to the next in an attempt to move to larger, better-paying firms; internal advancement is often impossible in such careers. In addition to the shift to services in the economy, the corresponding decline in firm size, as well as the decrease in firm size in manufacturing, has limited the development of career ladders. By their very nature, small firms and firms that are not growing offer few opportunities for internal advancement. It is easy to hit the ceiling in small firms.

Finally, two key institutions that have supported upward mobility and well-ordered advancement systems for large portions of the workforce—public sector employment and union contracts—are also in decline. In the public sector, civil service rules and requirements allowed for clear understanding of job access requirements and advancement possibilities. Likewise, in unionized settings, negotiation over seniority rules, wage structure, and upward bidding protocol encourages the development of clear internal labor markets. However, over the past 25 years private sector union density has fallen rapidly and now sits at just 10% of the labor force, less than one-third the union density at its peak in the United States. The public sector workforce has undergone massive reduction in size, due to both shrinking budgets for services and the privatization of public jobs, from garbage collection to social work. As these institutions have declined, more subjective systems for advancement, including personal connections and supervisor choice, may have become more important in labor market outcomes. Those who have little luck or few contacts to start with are then further marginalized in the labor market.

The Labor Market Results of Collapsing Ladders

Taken together, these trends result in less clarity in labor market signals, less training for incremental progress through the labor market, and fewer prospects for real careers for many workers. These effects, in turn, contribute to increased inequality in the labor market, less regularity in career trajectories, and greater influence of luck and connections on labor market positions. The result is a far more forbidding system for would-be labor market entrants. Job seekers can no longer "go down to the factory and sign up" with any confidence that they will have the skills needed for entry-level jobs or that those jobs will naturally put them on a career path of increasing income and security. The labor market does not provide sufficient information about transitions to work, transitions within work, and requirements for advancement. Many workers find themselves stuck in dead-end jobs or even in mid-level jobs without any obvious or clear route out of their situation. While these realities obviously cost workers—poverty and frustration are just part of the toll—they cost employers as well.

■ Dead-End Jobs

One of the results of the restructuring of the economy and of workplaces has been the emergence of a growing category of what are commonly known as dead-end jobs. Such jobs are characterized by very low wages and little hope for progression to jobs that pay better. Although it is clear that many, even most, workers make it out of dead-end jobs after a period of time, there is another segment of the labor force that "scuds" through dead-end jobs, never finding a way out. Workers with less education, women, people of color, and the young and old are disproportionately concentrated and trapped in dead-end jobs. As the service sector grows, and as firms increasingly contract for low-skill work (from janitorial work to data entry) from temp agencies and business service providers, some of the avenues that used to exist within firms for advancement out of low-skill jobs have been cut off. In these instances, entry-level employees cannot prove themselves and expect that proof to provide them with access to better-paying work in the firm; after all, these workers are not even formally employed by the firms in which they work.

However, dead-end jobs are not simply inevitable; in fact, they often produce a series of economic problems for both firms and workers. For example, firms are experiencing increasing problems with turnover in their lowest-skilled jobs.

Turnover produces significant costs in training and recruitment and reduces productivity and efficiency. As a result, firms are both reluctant and unable to train workers fully. One employer we spoke with described his hotel's need to reduce turnover from its current level of more than 100%: "That's probably our biggest problem right now—to get away from the transient-type employees, that this is just something to do for a couple of months and they're gone." At the same time, workers see few rewards for staying in these dead-end jobs, little training to improve their wages, and few opportunities for advancement. If workers were convinced that continuous tenure on the job would be rewarded through wage increases or advancement opportunities, they might choose to stay in their jobs. And if firms could be convinced that workers are more likely to stay on the job, they might be interested in investing more in the workforce.

■ Few Opportunities to Increase Skills

Further up in the labor market, at slightly higher wages and skill levels, the problem is repeated: Many workers, in jobs as diverse as machining and medical record coding, feel stuck in mid-level jobs. Workers in these jobs often do not have the time or inclination to invest in training outside work. On the employer side, few firms regularly upgrade the skills of their employees. Often, reaching the next rung on the ladder requires a prohibitive investment of time in schooling, such as certification in insurance billing, medical transcription, or skilled trades. In these jobs, systems for learning, skill upgrading, and training are largely overlooked or pursued by only a handful of employees.

Again, being stuck at mid-level work is a problem for both firms and employees. A highly trained workforce does increase the possible gains to firms, which are able to adjust to new demands in markets and new technologies of production. In fact, much of high-performance manufacturing requires a well-trained workforce, and firms in the service sector are also required to adapt to new technologies quickly and to adjust to new markets and changing demand. Indeed, a well-trained workforce can make a critical contribution to a firm's competitiveness. The problem is that a firm that wishes to make the transition to a high-performance workplace cannot create an appropriate system for training its workforce on its own. Many employers quite reasonably fear that their trained workers will be "poached" by other firms that can offer slightly higher wages without ever contributing to the original employer's investment. The fear of "free riders" pushes firms to invest less than they would like in training or to train more narrowly than would be optimal. These firms continue to pursue strategies that may require less of their workforces, promise less long-run productivity, and lower wages.

■ Informal Networks Get You In and Up

Asking, "How did you get your job?" often brings a common response: "My [friend/relation] told me about the opening and I applied." Clearly, not all jobs in the labor market are filled in this informal manner, but often, in the absence of reliable credentials or recommendations, employers turn to their own networks of employees and colleagues to help them fill open positions. At upper levels, the process may create glass ceilings, through which professional women and minorities cannot advance. Perhaps more important, at the bottom of the labor market, a reliance on informal networks may dig a bottomless pit for workers with less than a college degree, especially those who are women and/or minorities (Holzer 1996; Granovetter 1995). For a worker getting into his or her first job, a well-connected and working network of friends and relatives can help identify opportunities and ease access. As central-city residents have suffered increasing economic isolation, this first transition has become more difficult (Wilson 1987). Moreover, evidence of increasing labor market inequality within specific race and education groups (Danziger and Gottschalk 1993) also suggests that initial position in the labor market may be increasingly influential on future opportunity. That is, those who get off to a bad start in the labor market may find themselves further and further behind in part because they are not in a position to develop the sorts of contacts that can lead to better work.

The real problem with informal networks is that systems of access into the labor market, and the strategies for advancement through it (both within and across firms), are not clearly defined, identified, or understood. Obviously, information on the labor market—knowledge of opportunities, key actors, and critical institutions—can make all the difference in terms of outcomes. When the information is increasingly difficult to obtain, or when that information is more detailed for some than for others, inequality will be the result. To the extent that an individual's current position determines future opportunities, that inequality will increase over time. Our interviews with Wisconsin employers suggest that firms have little faith in verbal and written recommendations from previous supervisors and that they often prefer employee referrals. Further evidence of the negative effect of informal systems on marginalized workers comes from a multicity study of employers that shows that those who use tests to screen job applicants are much more likely to hire black candidates for jobs than are employers who rely on more subjective screens, such as interviews (Holzer 1996).

Given the strong anecdotal evidence suggesting that job displacement and career interruption are becoming ever more likely, the importance of establishing clarity in the labor market can only grow; churning in the market will require individuals to find new jobs more frequently. Without clearly identified pathways of access, churning will further damage the chances of those with little access to good informal sources of information.

Rebuilding Access and Advancement Systems

In sum, the result of these trends is a lack of clarity in the labor market. Workers are not aware of opportunities and have less systematic routes of access to and advancement through the labor market. The payoff to training and education in this context is unclear, and luck may have as much to do with labor market outcomes as skill. Overcoming these problems will require the development of institutional infrastructure that can provide the clarity, incremental skill growth, and career trajectories of the old corporate ladder system while still being flexible and responsive to emerging needs in the new economy. Improved labor market information (i.e., where demand is, what jobs pay, what skills are required) and more relevant training are essential elements of any response.

Improving labor market navigation requires the development of a clear map of the labor market and improved supports for navigating through it. Such a map would help individuals overcome the barrier of informal access to work by establishing a clear system of rules on access and advance. Moreover, the information would help both students and employers plan for the future. One firm representative we interviewed summarized it this way: "As these students leave high school, it would be nice if there was something to say: 'This is what the need is in your community. There's this number of jobs that are becoming available in the community in this area, and these are the skills that you're going to need to fill one of those positions.' . . . And I don't know if that's something that's being done now." A labor market map would also allow adults opportunities to change careers, and would aid in the process of finding new employment after job loss.

Entry-level and incumbent workers need to see a way up through the labor market. Employers need more reliable systems for identifying, hiring, training, and promoting workers. Systems can be developed to meet these needs simultaneously, but to work, they need a focus. Our own experience in Wisconsin suggests that leading sectors in regional labor markets are a powerful means of providing this sort of focus. Moreover, these sectoral efforts can serve the additional function of rationalizing and better directing public institutions that need better information about employer needs and skill demands to improve their efficiency. We return to public administration issues at the end of this chapter; first, we will describe sectoral projects in both theory and practice.

Sectoral Projects to Improve Opportunities in the New Economy

Our work in Wisconsin has focused on the development of sectoral intermediaries that provide solutions to industry problems and use those solutions to improve training for incumbent workers and increase access to the industry of disadvantaged workers. These projects and many others are known as *sectoral strategies*. They have excited considerable interest in recent years as the "natural" workings of the labor market have soured for many Americans, and as growing recognition of funding and information constraints on government have undermined confidence in the public sector's ability to turn that around.

Experience from the field suggests that these projects can simultaneously advance the interests of sectors and low-income communities by developing trusted sectoral intermediaries that respond reliably to private sector needs. These intermediaries are close to demand in the labor market and can leverage that connection to produce improved low-income access to jobs and training on jobs in the sector. The success of these strategies must rely on the sectoral intermediary's ability to provide firms or the sector at large with better services and programs than are available without the intermediary, or firms will not be involved. Projects, programs, focus, and success in the field are diverse, but the need to lead the industry, leverage funds within the industry, and develop efficiencies are all keys to the success of the initiatives.

Sectoral Initiatives in Theory

How are positive effects produced—in effect, creating a "win-win" situation for firms, workers, and new labor market entrants from the community? By focusing such strategies on sectors, a collection of firms with shared production methods and/or labor forces, these projects are supported by and rely on three key efficiencies. First, there are considerable economies of scale to be realized in working with a cross section of firms with shared labor force needs. Unlike workforce development or job connection initiatives that have adopted a narrowly customized firm-by-firm approach, the sectoral approach does away with the need for constantly renewing personal relationships and inspiring management goodwill or civic-mindedness. Through the organization of a sector, solutions to recurrent problems can be improved and programs refined. The learning that results improves efficiency.

Second, a sectoral initiative can leverage economies of scope. A sectoral intermediary that understands and responds to the shared needs of a number of firms can develop diverse programs that will weather economic shifts and the problems of single firms. Just as the scale of the project can be used to improve continuously the common elements of the curriculum and training routines, the scope of the project protects it from being captured by a single firm and becoming uniquely beholden to that firm's needs and fortunes.

Finally, a successful sectoral initiative will develop positive network externalities, which are generated as firms come together and find ways to solve common problems. If initiative programs are helping multiple firms in a sector, then the region's sector at large can become more vibrant and competitive. Firms, by sharing information and solutions, also may become more tied into the local community. So, as a valued sectoral intermediary is developed in a region, the sector can become revitalized and more rooted in the local economy.

Sounds Nice, But Can It Really Work? Sectoral Initiatives in Practice

At the Center on Wisconsin Strategy (COWS), a policy and research center at the University of Wisconsin–Madison, three of our major projects are focused on developing the sectoral structures that can contribute to workers' opportunities. The first is the Wisconsin Regional Training Partnership (WRTP), a relatively mature initiative in Milwaukee's metalworking sector. The second is the Milwaukee Jobs Initiative (MJI), which brings a much greater focus on connecting disadvantaged workers to jobs in specific high-wage and increasingly organized industries. The third is the infant Community Career Ladder Project in the metropolitan Madison area. Although not perfect, these efforts have achieved some success, and from our experience working on them we can offer some important insights on the promise, design, and development of sectoral strategies in general.

■ Wisconsin Regional Training Partnership

Based in the metalworking industry, the WRTP has a membership of more than 40 firms collectively employing approximately 60,000 workers in southeastern Wisconsin. Founded in 1992, it has grown to be the largest sectoral training consortium in the country, and the most advanced in overall program goals. Under the terms of its joint labor-management governance, the WRTP requires that member firms do the following: benchmark a growing percentage of payroll to training frontline workers, train according to standards set on a suprafirm basis, gear hiring and internal labor market promotions to worker achievement on those standards, and administer the enhanced training budgets (resulting from their benchmarked contributions) through joint labor-management committees.

Over the past 5 years, the WRTP developed programming in three major areas of activity: incumbent worker training, modernization, and future workforce development. In regard to the first, new technologies and new work organization require workers with new skills. For many workers, the transition to a "continuous innovation" environment is impossible without considerable training. WRTP members have prioritized activity that can aid firms as they seek to develop and/or improve their workplace education centers. The WRTP provides assistance as firms develop new centers. WRTP staff direct firms to external resources available for workplace skills centers, such as funding sources and curriculum developers; they also work inside firms to help develop the labor-management collaboration on which any successful skills center relies. Without labor collaboration on the project, the skills taught at the center can easily be irrelevant to worker needs, thus resulting in shop-floor skill gaps. Without the context of workforce buy-in and contribution to the development of the skills center, the significant investment in a workplace skills center can be wasted. The WRTP facilitates cross-site and cross-union learning about workplace skills centers and has also helped a series of small shops to develop jointly operated centers. Finally, the WRTP helps firms and unions develop workplace awareness of the centers through peer adviser networks.

Competitive pressures in the metalworking industry require that firms have access to and adopt rapidly advancing technology. In response to this industry need, the WRTP focuses considerable attention on modernization of member firms. Often, new technology and new work organization go hand in hand. Many firms, especially the WRTP's smaller member firms, do not have sufficient resources to commit to their own modernization. The state of Wisconsin has developed the Manufacturing Extension Program (MEP) to assist firms as they identify, adopt, and adjust to needed technological modernization. The WRTP is collaborating closely with the state MEP to ensure that member firms have access to the resources of the MEP and that those resources will help to serve member firms. Again, collaboration among member firms can improve learning and knowledge of technologies and the challenges that come with modernization. Both management and labor union members can investigate options and discuss the effects of modernization with firms that have already adopted new technologies. This shared experience can ease the process of modernization and improve the efficiency with which new technologies are adopted by allowing firms to avoid common mistakes.

Finally, regarding future workforce needs, WRTP programming is directed toward both school-to-work initiatives and programs for dislocated and disadvantaged workers. In the school-to-work area, WRTP staff are working with the public schools to develop a better connection between students and area manufacturers. Member firms have committed to a series of "take the teacher to work" days, involving 300 Milwaukee public school teachers, and to programs for youth apprentices and co-op students. Given the high level of employer involvement, the effort promises to move several hundred young people into manufacturing youth apprenticeship in the next few years. Although still small compared with European apprenticeship programs, this WRTP program dwarfs the expected participation in other youth apprenticeship programs in Wisconsin. We discuss the WRTP's collaboration on the Milwaukee Jobs Initiative in greater detail below.

The apparent results of the WRTP have included significant improvement in the skill levels of the workforce, stabilization of employment in this hard-hit and highly competitive industry, wage improvements for incumbent workers, clear markers for entry-level and incumbent workers regarding job expectations and career advancement, and considerable improvement in the general quality of labor-management relations in affected firms. Although the work of the WRTP is often customized to meet specific firm needs, the consistent attention to training and modernization issues throughout the sector develops skill benchmarks that improve workers' mobility across firms, not only within them.

■ Milwaukee Jobs Initiative

The Milwaukee Jobs Initiative is a 7-year project to improve the systems that connect central-city residents to jobs. Focusing in the first year on manufacturing, printing, and construction jobs, MJI projects will connect at least 240 unemployed or underemployed central-city residents to family-supporting work. In each of the three targeted sectors, the MJI will improve the organization, integration, and coordination of actors on both the demand and supply sides of the labor market. The MJI builds on the experience of the WRTP and seeks to develop stronger links between the community and organized sectors.

In manufacturing, the WRTP is working with the MJI to involve a broad range of firms in a future workforce project that will systematize the links between firms and central-city residents. For this project, the WRTP is identifying and recruiting manufacturing firms that will be hiring in the near future, working with these firms to identify the exact skills needed for entry-level employment, working with the firms and training providers to develop curricula that will prepare workers for these jobs, and working inside firms to develop the sorts of peer advising systems that can help support new labor market entrants from inside the plant. On the community side, the project enlists community-based organizations to recruit, screen, and support prospective program participants. Given the low area unemployment rate, employer response to the project has been extremely positive. Given the high wages offered in WRTP firms, generally around $10.00 per hour to start and with good benefits, community interest in the project is also high. This "jobs connection" was made for more than 100 central-city Milwaukee residents in 1997, and is projected to grow rapidly over the next few years.

The WRTP focused originally on incumbent worker training systems, and these systems provide the foundation for the MJI project. First, project participants will be entering firms with some basic understanding of work in manufacturing, but as training is short-term (usually 12-18 weeks), workers will need to develop skills on the job. The incumbent worker training systems provide this necessary worker training infrastructure so that the entry-level workers can develop more advanced skills. Second, the peer adviser networks that were developed to encourage workers to use workplace education centers provide a natural infrastructure of support for new shop workers. Finally, providing training opportunities to incumbent workers also helps create an environment that is more accepting of new workers. Without workplace innovation requiring higher skills, training for new entrants and their insertion into low-skill environments would only frustrate and cause resentment among incumbent workers.

In addition to the manufacturing project, the MJI will invest in projects that connect central-city residents to jobs in printing and construction. The focus of these projects, as in the manufacturing project, is to organize on both the demand and supply sides of the labor market. On the demand side, projects will help develop both retention and incumbent worker training systems. On the supply

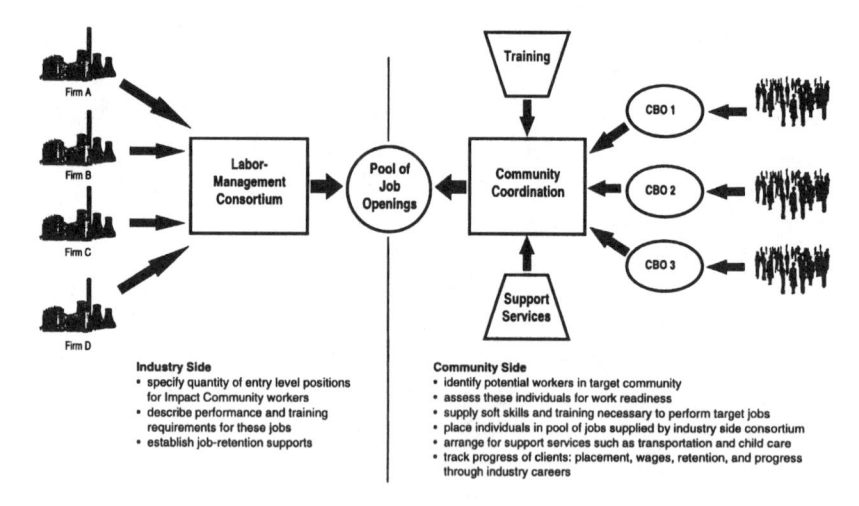

Figure 4.1. Organizing Job Connections on the Supply and Demand Sides of the Labor Market

side, projects will encourage coordination of community resources for recruiting and supporting new labor market entrants. Figure 4.1 displays the general structure of MJI projects in manufacturing, printing, and construction.

■ Community Career Ladder Project

In 1995, the Dane County (home of Madison, Wisconsin) Board of Supervisors reconstituted the Economic Summit Council—a blue-ribbon commission comprising leaders from Dane County business, labor, public, and nonprofit sectors. The Summit Council is charged with developing a strategic vision for economic and workforce development in Dane County. One element of this plan is the Community Career Ladder Project, designed to make "jobs with a future" available to all Dane County residents. COWS has served as the leader in technical assistance and design of the Community Career Ladders Project.

In reviewing our experience with the WRTP, best practices from around the country, and detailed quantitative and qualitative analysis of local labor market conditions and needs, we identified three key sectors for development of sectoral intermediaries that could support high-performance work organization on an industrywide basis. In three sectors—manufacturing, health care, and insurance and finance—COWS conducted extensive interviews with human resource and training personnel at leading firms to identify the skill and workforce needs shared by the firms in each industry. This process recently led to employer-union decisions in each of the sectors to establish a sectoral consortium to address these needs.

In manufacturing, COWS found that many local manufacturers are experimenting with work reorganization in response to product-market demands for increased quality, flexibility, and innovation, coupled with lower costs. Most firms were also investing heavily in new equipment, much of it computer controlled, and trying to stay close to the leading edge of technology in their sector. The combined effect of work reorganization and computerized technologies has been to raise sharply the basic skills demanded of incumbent workers. Requirements for more advanced process improvement and teamworking skills have also grown significantly. One of the most striking features of the new skill needs is that they are neither firm nor sector specific and have wide applications beyond metalworking itself. The shared industry problems can be broken into four main areas: recruiting and retaining entry-level workers, upgrading the skills of incumbent employees, managing the challenges of work reorganization, and meeting future workforce needs. The activities of the sectoral consortium will concentrate on solving these problems.

Our study found the health care industry to be in rapid flux, with restructuring and cost-containment efforts altering the structure of work. Shared problems emerged within the major industry segments. Hospitals have difficulty retaining entry-level workers in their food service and housekeeping jobs, and they have a hard time filling specific professional and paraprofessional positions. Health maintenance organizations have the problems of an emerging industry—courses and training are designed for office rather than clinic settings. Extended care and home health care providers are struggling to maintain quality care in a tight financial situation; reimbursement for services has simply not kept pace with the costs of care. Although the survey shows distinct problems for the industry segments, it also indicates that some of the pressing problems of the industry—occupational shortages, retention problems, and needs to increase the skills of the workforce—will be solved efficiently only through collaborative industry-wide approaches. The efforts of the sectoral consortium will focus on reducing turnover in lower-end jobs through certification and job pairing; selected industrywide programs of training, on an occupational basis, to meet special employment needs; and an attempt in at least one major facility to replicate a career ladder program developed in Cape Cod Hospital.

In the insurance and finance sector, we suggested the development of a program to move nontechnical (especially clerical) workers into more technical positions while meeting firms' needs for information systems (IS) analysts and programmers. In insurance in particular, technological change has had major effects on the workforce, in terms of both workforce structure and skill needs. Nontechnical staff are increasingly required to upgrade their skills to meet the requirements of improved technology. Perhaps most significant, all firms that have IS departments are experiencing a dramatic shortage of qualified personnel. The consortium now formed will concentrate first on developing a modularized IS curriculum and will create a "programmer trainee" position. This will allow firms to send their clerical staff through IS training in such a way that the skills created will bring immediate benefits to the firms.

In each of these sectors, the barriers to industry collaboration are significant. Firms in each industry are often in direct competition for customers and workers. Moreover, their problems, although similar, are not always exactly the same; involvement in the collaboration requires firms to work less on narrowly customized solutions and more on collective, but slightly less direct, solutions to problems. However, industry reaction to the sector studies and proposals for creation of cooperative solutions to shared problems has been very positive. Firms see many advantages to partnerships focused on training and modernization, and, given good understanding of local issues and some sense of possible joint solutions, they are willing to work in new ways.

What do leaders in these industries hope to gain from working together? First, by formalizing collaborative work on skill and training issues, firms and unions improve their ability to learn from each other. The connection on these issues provides a forum for best practice and helps to keep firms from constantly needing to reinvent the wheel. Second, working together, firms can also develop a more unified industry voice. Speaking more clearly as an industry, firms can then engage education and training providers and the broader public in an informed dialogue on industry needs and opportunities. Finally, only by working together can firms take advantage of the economies of scale provided by coordination. At least for now, labor and management both seem to believe that the potential benefits of the project will outweigh the costs.

Building Blocks for a New Public Sector Agenda

In the short term, these evolving sectoral intermediaries will play a critical role in improving incumbent workers' advancement in and disadvantaged workers' access to careers in the industry. The experience of the WRTP also shows that as the partnership grows in scale and scope, it is increasingly well positioned to provide public institutions with representative and detailed information on industry needs. If multiple consortia were organized in a regional leading industries, the efficiency and efficacy of publicly administered labor market services would improve dramatically. That conclusion is supported by a consideration of the principal problems in labor market administration.

Problems in Labor Market Administration

The nation's myriad jobs support systems—training, education, assessment, transportation, and child care, to name just a few—are riddled with inadequacies. At the federal level, despite some success in consolidation, national training efforts remain littered across more than 100 different programs, administered by 14 different federal departments and agencies—many with identical target populations and purposes, but all with different mandates. This sprawling effort becomes slightly less intelligible at the state level, where principal responsibility for federal program administration and control over a welter of additional programs resides. It is almost unnavigable at the local level, where much effective authority over postsecondary training institutions (including most community colleges), as well as local school districts, is exerted. The maze of programs and decentralized administration has grown into being over more than a century of political conflict over the appropriate role of government in human capital formation, and its structure reflects the best and worst aspects of federalism—from the room it provides for efficient local servicing and experiment to the inequalities it breeds in the byzantine complexities of state and local tax codes. Each program has a constituency, and systemic reform is not aided by the reserved—and now growing—powers of state and local governments.

As a result, training institutions are not teaching workers the skills that employers need. Education programs do not effectively situate and motivate adult learning, and students become dissatisfied or apathetic. The assessment tools used in a community in referring individuals to the same programs can be widely different in form and validity.

In part, these problems are the result of financial constraints—increasingly evident in an age of taxpayer reluctance, itself driven by declining incomes. In part, they result from institutional disadvantages—the fact that government is and always has been inept at capturing the information and local knowledge key to economic decision making, particularly inside the firm. The combination is commonly lethal, in any case, to effective intervention, with government stumbling after what it takes to be the latest industry "trend" without the resources to drive it, and almost always finding that by the time it arrives on the scene the underlying economic conditions have changed.

Fundamentally, the public sector lacks the resources required to provide the level of training that could actually contribute to workers' and firms' economic performance. Moreover, the public sector lacks clear access to private sector information that could more efficiently direct the resources available in the public sector. Perhaps most important, there is very little attention by public sector training providers to either supply or demand in the labor market. Because there is so little simultaneous, coordinated attention to economic development, job quality, and workforce development, these areas of policy often run at cross-purposes: One agency supports economic development without regard to job quality, while another agency seeks to improve specific skills of the workforce without attention to economic development priorities.

In response to these information and resource problems, public sector work has often become highly customized, personalized, and nonrepresentative. The customization is apparent when training institutions develop programs around the narrowly defined needs of a single firm and work on a firm-by-firm basis. Public sector work is highly personalized in job centers where counselors rely on personal connections with firm human resources managers to get their clients into jobs. Advisory committee structures generally rely on civic-mindedness or narrowly interested business representatives. The result is a fractured system with multiple entry points, bureaucratic rules, conflicting priorities, inconsistent and nonrepresentative employer participation, poor coordination, and missed opportunities for labor market upgrading. In spite of the real progress one-stop job centers have made in overcoming some of these problems, they are still significant.

Sectoral partnerships could play a valuable role in overcoming these public sector problems. In effect, they provide at the regional level what is not provided nationally—a genuine infrastructure of industry and union collaboration that both drives industries toward the need for more demanding skills and provides the flow of information, and assurances against free riding, needed to meet it. They could provide private sector-driven information about the market and emerging needs in the labor market, which would permit more intelligent programming to meet those needs. Additionally, sectoral initiatives could leverage firm investments in training, workplace reorganization, and modernization. Well-designed sectoral partnerships in leading industries are uniquely positioned to provide the public sector with the "inside" information on industry needs that is representative, self-interested, and at scale. With strong sectoral

```
                    ┌─────────────────────────────┐
                    │  Regional Labor Market Board │
                    └─────────────────────────────┘
         ┌───────────────────────┼───────────────────────┐
┌──────────────────┐  ┌──────────────────┐  ┌──────────────────┐
│ Sectoral Consortium│ │Sectoral Consortium│ │Sectoral Consortium│
│   Representative   │ │  Representative   │ │  Representative   │
│  ex., Hospitality  │ │   ex., Health     │ │ ex., Manufacturing│
└──────────────────┘  └──────────────────┘  └──────────────────┘
```

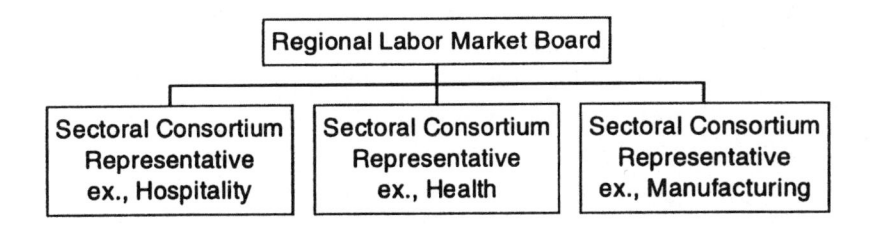

Figure 4.2. Sectoral Organization as the Basis for Regional Labor Market Administration

NOTE: Representatives of consortia serve on the regional labor market board to ensure that public sector resources are well directed. Public sector representatives (from school districts, job centers, technical colleges, manufacturing firms, and so on) also sit on the board. This type of organization provides a coordinated public-private interface on regional labor market priorities and needs.

partnerships, public agencies and community-based organizations have the sort of one-stop demand-side coordination that job centers provide on the supply side of the labor market. School-to-work programs, technical colleges, community training programs, and support services can relate to the industry in one forum. Given pressures for devolution, moreover, there is no reason such efforts could not be more effectively integrated into public labor market administration.

If all of a region's most important sectors were organized into consortia to work on workforce development and modernization issues, then representatives from those organizations could sit together on a regional labor market board that could direct public sector resources devoted to training and education, and human services funding devoted to workforce development and job connections (see Figure 4.2). This regional labor market board would be representative of the regional economy, and members of the board would be responsible to their sectoral organization rather than just to a single firm. Additionally, public sector systems from schools to welfare departments to technical training institutions could receive accurate information and projections of opportunity and skill requirements for the workforce.

In an age of growing inequality and central-city economic isolation, devolution appears as yet another attack on the already weak public systems that are responsible for responding to the needs of central-city residents. At the same time, devolution presents an opportunity to press for real systems reform in local labor markets. The organization of lead sectors can contribute positively to that rationalization process. Regional labor market boards could be the sites where community, employer, and public sector needs are negotiated and responded to in a flexible, representative, and responsive manner. Our experience in Wisconsin suggests that such organizations can be developed; that private, public, and community players are ready to make the necessary changes; and that the positive effects for everyone involved will far outweigh the difficulties in bringing them about.

References

Danziger, S., and P. Gottschalk. 1993. *Uneven tides: Rising inequality in America.* New York: Russell Sage Foundation.

Granovetter, M. 1995. *Getting a job: A study of contacts and careers.* Chicago: University of Chicago Press.

Holzer, H. J. 1996. *What employers want: Job prospects for less-educated workers.* New York: Russell Sage Foundation.

Wilson, W. J. 1987. *The truly disadvantaged: The inner city, the underclass, and public policy.* Chicago: University of Chicago Press.

PART II

TECHNIQUES

5

Regional Economic Analysis to Support Job Development Strategies

BRIAN BOSWORTH

Working on the Demand Side of the Labor Market

One of the more frequent criticisms of conventional public sector employment and training programs is that they tend to be narrowly focused on the supply side of the labor market. Supply-side-dominated strategies end up poorly serving individuals who look for career opportunities, not just low-skill, low-wage jobs; firms that seek to gain competitive advantage through worker skills; and regional economies that seek more wealth creation.

A job development strategy that incorporates a demand-side approach would get employers more involved in helping to shape training programs. It would help firms understand how to develop and deepen skills through improved training practices, better forms of work organization, and other human resource development practices. It would support employers in setting skill standards, assessing the performance of both new entrants and current workers, credentialing important skill thresholds and achievements, and facilitating career mobility. A strategy that incorporates demand-side concerns would seek to optimize returns to the regional economy by focusing its services, to the extent consistent with the objective of helping job seekers get good jobs, on those firms where a more highly skilled workforce might most significantly contribute to aggregate wealth in the region.

Doing this effectively, however, requires a deeper understanding of the regional economy than many job development organizations have attempted. As job development practitioners seek to influence demand, to build systems that move people from low-wage jobs to better ones, and to use work to pull skill formation, they must be able to identify those firms, business sectors, and clusters most critical to the region's economy. They ought to know how to help organize these firms to pursue collective solutions to common human resource problems, building relationships with groups of firms in ways that will enhance the chances of influencing their business and human resources strategies.

Good programs already do this. In fact, what stands out from a review of the more successful programs is their relative success in becoming integrated into the internal labor market of firms and their industry associations (Clark and Dawson 1995; Rosenfeld 1995a, 1995b; Osterman 1988; Harrison, Weiss, and Gant 1995; see also Giloth, Chapter 1, this volume).

Of course, some employers have little interest in changing their human resource practices in directions that would pull higher skills out of their current workers and prospective employees. Some prefer a "low road" to competitive success, and although they may occasionally lament the shortage of skilled workers, they are not likely to invest in new systems that contribute to skill formation. What is important is for programs to work with the firms that have opted for the "high road" of high-value-added goods and services, who are willing to pay high wages for high skills and who see a responsibility to help fashion new strategies of skill development. What is critical, of course, is to figure out which firms those are, what might make a difference to them, and how to get them involved.

Regional Economic Analysis to Support Demand-Side Job Development Strategies

Regional economic analysis that will support effective demand-side strategies of employment and training demands a good deal of knowledge about the firms in those systems: how they are connected to each other in the regional economy, what factors drive their competitiveness, and what the points of leverage are where thoughtful public policy and collective private sector actions can affect both their competitiveness and their human resource policies.

This chapter provides broad and selective sequential guidance for practitioners interested in analyzing regional economies: first by developing basic familiarity with the structure, composition, and performance of the regional economy; then by investigating the special needs of major industry groups or sectors; and finally by focusing on the major intersectoral business clusters that drive the regional economy. This approach should have some immediate payoffs in allowing a researcher to identify quickly those specific companies in the region that are adding new employees and to initiate structured conversations with them about human resource practices, current and future employment needs, and the relationship between entry-level skills and advancement to jobs of greater responsibility and better pay. However, supporting the long-term competitiveness needs of employers and trying to influence their human resource strategies requires a clear understanding of which particular industries enjoy competitive advantage in the region and where further gains in worker productivity would contribute significant wealth to the region.

Sector analysis permits examination of how changes in these most important industries are influencing their labor market behavior and needs. It leads to an assessment of their collective workforce needs and supports them as they pursue joint solutions to common human resource problems. Cluster analysis investigates the relationships among interdependent industry groups whose fortunes are intertwined in the regional economy. It suggests how to probe the current and prospective performance of these clusters and how to identify programs and policies that will both strengthen the occupational groupings that support these clusters and, in the process, reduce barriers to the rapid diffusion of knowledge and innovation among firms in the clusters, helping them find further economic advantage through their relationships with each other.

At all three levels of investigation discussed here, economic analysis is not just "running the numbers." Statistical analysis is merely a place to begin; by itself, it will reveal very little about a dynamic regional economy. The more critical tasks are those involving structured dialogue with the key firms in a region. Employment and human resource development practices, especially in the small-firm sector of the economy that provides most entry-level opportunities, are very informal and unorganized. Understanding them well enough to help influence them requires a sustained relationship based on trust and the expectation of reciprocity.

Getting Started: Developing a Quick Profile of the Regional Economy

For most workforce training programs, developing an understanding of the regional economy begins with some pretty basic steps: defining the region, developing a profile of the structure and composition of enterprise, and starting a dialogue with employers about their human resource issues and skill requirements. These steps build the foundation for more sophisticated analysis.

Selecting the Geographic Scope of Analysis

Most job development programs, because they are supply-side oriented, define the geographic scope of their work based on where the residents of the target community might reasonably expect to commute to work. But the first lesson in working on the demand side is that this geography must be based on relationships among firms. Influencing the employment practices of firms located within close commuting distance of the job seekers requires a scope of analysis and intervention that is much wider than commuting patterns; the close-in firms may be very dependent on the firms further out. What is needed is a strategy that deals with the metropolitan economy as a whole, as an interdependent system of firms and workers tied to each other by complex business and employment relationships. This means analyzing the economy on the basis of the economic region—the geographic area that contains the majority of the transactions that take place among firms whose behavior might be influenced.

For job development purposes, that economic region is most appropriately understood to be a metropolitan statistical area (MSA). MSAs are defined periodically for federal statistical use by the Office of Management and Budget. The boundaries are always coincidental with the boundaries of counties, but most MSAs consist of more than one county. Typically, an MSA includes a large population nucleus and certain counties around that nucleus according to relative measures of population density and commuting patterns.[1]

Assembling Secondary Data

The most logical way to begin more detailed analysis of the regional economy is with the collection and organization of easily accessible economic data from secondary sources about firms, employment, and payroll. The basis of this is the Standard Industrial Classification (SIC) system of the U.S. Bureau of the Census, which classifies all business establishments based on the products or services they sell.[2]

The SIC system is less precise than it might at first appear. The basis for classification is the *primary* product of the establishment, and this sometimes masks important economic activity. Also, modern production and information-processing technologies give firms much more flexibility than in years past to shift quickly from one product mix to another. SIC codes may exaggerate the boundaries separating firms and disguise complementarities. This means that although SIC categories offer the first logical step in an assessment of the composition of a regional economy, the classifications must be supported by more qualitative research.

A researcher might most appropriately begin with preparation of a table that shows for each major industry group (the SIC code at the two-digit level) the number of establishments, total employment, average annual payroll per employee, value added per employee (for manufacturing), and receipts per employee (for services). Developing this information for the key industries in the region will help the researcher to locate the major sources of change in the regional economy and to establish which are the major industry groups and the industry groups that are gaining or losing within the regional economy. This helps to highlight opportunities for quick-hit involvement with those groups that appear to be growing most rapidly and generating employment demand. Most of the firm data are available (with a 3-year time lag) from the annual publication *County Business Patterns*.[3] Ideally, the researcher would pull this information together for 3- to 5-year intervals over the past 12-15 years, to get a peek at any trends that have been developing over time. It is frustrating that the most recent data in *County Business Patterns* and U.S. Census publications are at least 3 years old. Certainly most regional economies see important changes within such a period of time. Still, the differences are not likely to be dramatic for the purposes of this summary analysis, and more up-to-date information can be developed when it comes to identifying the real companies behind the numbers.

Once the SIC analysis demonstrates which industries seem to be generating the most employment growth and wealth in the region, the researcher needs to assess the wage levels generally paid for those occupations in the region and to determine what employers expect in the way of skills and experience. The Bureau of Labor Statistics (BLS) publishes a series of reports that can provide a running start on these wage issues.[4]

Giving Priority Attention to Value-Adding Enterprises

Supply-side approaches to job development usually operate on the premise that a job is a job, and it is just as legitimate to prepare workers for entry-level jobs in retail establishments as in manufacturing firms. A strategy that incorporates demand-side concerns would be concerned with at least understanding which kinds of enterprises are contributing the greatest amount to the regional economy. Examining value added per employee is a way to gain this appreciation.

Value added is a term used commonly in manufacturing, but it is also valid when applied to nonmanufacturing firms as well. Put most simply, it is the difference between what the firm pays for the raw material, parts, and components it buys and the amount for which it sells its finished goods. The more the firm adds value on a per employee basis, the more it creates wealth and the greater the economic return to workers, managers, and investors. Value can be added through the application of sophisticated process technologies and the smarter work of highly skilled workers and managers. Value can be added in the design and engineering process through the incorporation of product features and capabilities for which the customer is willing to pay a premium. Value can be added in delivery logistics through the ability to get the product to the customer in the time, sequence, and mix that the customer needs. Value added is therefore not strictly a matter of productivity; it also reflects quality and service.

Those firms using relatively basic technologies, low-skilled workers and managers, and very traditional business practices are usually not going to be able to add a great deal of value to the raw materials and component goods with which they start. They will be able to compete only at the relatively low ends of their markets, where products must meet far less exacting requirements than at the high ends, where they are less differentiated, and, therefore, where they tend to compete chiefly on the basis of price. On the other hand, those firms that figure out more quickly than others how to adopt or adapt the highest levels of technologies, how to deploy very highly skilled workers into organizational systems that impel learning, and how to use business practices that promote customer response and quality products will be able to compete more aggressively at the highest ends of their markets.

Any region would want an economy of firms that are very good at producing the highest-value goods and services. These firms pay high wages and contribute to a rising standard of living for residents of their regions. If firms are slow in gaining the ability to produce high-value goods and therefore are forced to compete at lower levels of value added, they must compete on the dimension of price rather than value, paying lower rather than higher wages. They will gradually migrate to low-cost production environments or they will exert downward pressure on production costs where they are, dragging the overall standard of living down in the process.

Figures on the average value added per employee by industry groups are available from the 5-year *Census of Manufactures*. For service firms such figures are not as easily available, but receipts per employee may be used as an indicator, and these data are available from the *Census of Service Industries*. These reports are available directly from the Department of Commerce or may be found at good local libraries.[5]

Getting Names to Go With the Numbers and Doing Primary Research

The quick analysis suggested above yields only numbers. The next and much more critical step for the researcher is to decide which firms to talk to. Among the best sources of information on specific firms are the publications of proprietary information companies such as Harris Information Services and Dun & Bradstreet Information Services. These can provide, for any SIC code, by postal zip (or county or MSA), the names of all the companies, their addresses, parents if any, actual employment figures, estimated annual sales, and contact names and phone numbers. Using these or similar data sources, the researcher can identify most of the firms in a region that are members of the major industry groups or industry groups that appear to be growing most rapidly, and then employ these data to generate mailing and call lists for interviews or surveys of employment practices. These files can even be imported onto commercially available geographic mapping software that will show the locations of the firms in the metro area.

This list can serve as the basis for primary research, for asking a limited set of companies about the details of their hiring practices and trying to pinpoint particular barriers and opportunities for employment. Table 5.1 provides examples of some of the questions the researcher might pose to the representatives of these firms.

While collecting this information, the researcher may also find it productive to organize focus groups to present, validate, and add texture to the apparent findings. Such focus groups can also be a very useful way to find champions within the industry who are willing to get involved in the design and manage-

TABLE 5.1 Interview Questions for Firms

What are the occupations for which you hire entry-level workers?
Where do your new hires come from?
What wages do you pay for entry level in those occupations?
What nonwage benefits are provided to new hires and when?
What wage increments are generally available at various intervals of job tenure?
What percentage of entry-level employees can be expected to progress through these various levels?
What is the highest level of job classification and wages that might reasonably be expected in this occupational path?
What educational level do you generally expect for these entry-level positions?
What special training is required?
Do you train on the job? How much? How?
How many employees from this community have you hired in past years?
Do you see any special barriers to hiring good candidates from this community?
What are your suggestions for improving the job readiness and career advancement of candidates from this community?
What kinds of on-the-job support would be important to you in considering the employment of individuals with poor work histories and poorly developed skills?
Do you meet a major share of your employment needs through temporary help agencies? How much in terms of full-time equivalencies? Will this increase or decrease?
Are you satisfied that the education, job training, and labor market systems prevailing in the region work well for your firm and will continue to meet your needs over time?
Would you be willing to help in the design of better ways to accelerate job readiness?

ment of programs. This can provide the firms with a clear path toward more in-depth participation.

The Second Level of Research: Sector Analysis

Moving beyond quick-start strategies to sectoral initiatives requires the researcher to conduct more systematic analysis of key business groupings and of the factors that will drive employment decisions in those groupings. This means concentrating on sectors where regional advantage will contribute to long-term growth.

At any particular point, an entrepreneur may choose to establish a company in a particular place for any one of several idiosyncratic reasons. For a number of other equally idiosyncratic reasons, that firm may do well for a time, even for a long time. But when a significant number of similar firms prosper over time in a particular region, compared with other regions, that is not mere chance; something is giving that set of firms regional advantage. It may be proximity to raw materials, markets, specialized labor, essential infrastructure, world-class research, sympathetic financiers, or some combination of these. It may be that spin-outs from a particular company have over several years created a specialized business culture that has gained sustaining momentum and synergy even if the original advantages have evaporated. Obviously, regional advantage can be volatile; significant changes in technology or markets can erode advantage, and in the modern economy this can happen quickly. So in trying to identify regional advantage, a researcher needs to look carefully for changes over time and hedge his or her bets by working with several different industry groups.

In examining sector advantage, one of the first things a researcher should do is compare the structure of the regional economy with that of the national economy (or of the state, the multistate census region, or other regions) to reveal, first, where there are particular sectoral concentrations of economic activity by SIC code in the region relative to the comparison, and second, where the trends of change in those regional concentrations—especially in terms of employment, number of establishments, and average payroll—differ from trends in the comparison area.

This secondary data collection and analysis will reveal those business groupings where the region appears to be gaining or losing some comparative advantage. Both will be important topics for further secondary and primary research. In those groupings where the location quotient is declining, there is probably some erosion of what has been a source of strength in the regional economy.[6] Shift-share analysis will reveal if that decline or increase is faster than for the nation as a whole.[7] The development of an understanding of industry dynamics requires the researcher to draw upon a very eclectic mix of sources. There are some U.S. government reports that can be helpful. The U.S. Department of Commerce publishes the annual *Industrial Outlook,* which summarizes recent data on major trends in several hundred industries.[8] Trade associations frequently publish reports on key issues facing the industry, and trade periodicals can be a rich source of inside information about the industry in question.

Careful sector analysis also includes inspection of the spatial dimensions of these changes, especially those dealing with employment. Armed with good firm-specific data from the proprietary information sources, a researcher may be able to compare and contrast these groups by postal zip code. Reliance only on data that are presented essentially at the MSA level may not reveal any shifting of jobs from one part of the region to another. These intraregional changes are especially important in their influence on population groups whose mobility is limited by lack of transportation and sharply constrained employment networks.

Organizing this secondary information is mostly just the researcher's preparation for primary research—contacting the firms directly. Face-to-face interviews are the most productive form of contact, but may be prohibitive at this stage because of the number of companies involved. Telephone interviews may not be satisfactory, because the issues that need to be pursued are very qualitative and thus hard to cover in phone conversation. The researcher might find group interviews to be a good way to get started; he or she might then distribute short surveys to gather qualitative information to confirm issues identified in the group interviews. Among the questions the researcher would want to ask during this primary research are any that might lead to discussion that could reveal important trends, competitiveness barriers and opportunities, and special considerations that will affect labor market demand over the next several years.

Using Sector Analysis to Shape Strategy

Employment and training programs that focus on a major industry group or industry groups and target an occupation or group of occupations common to most firms in those groups represent a huge advance over more conventional job training and placement programs, which often have as their target individual firms with unrelated occupations. Building new mechanisms for interfirm cooperation in pursuing joint solutions to common human resource development problems provides a far stronger basis for engaging employers in training programs and in influencing the quality and quantity of their demand for workers. These strategies are consistent with trends in the economic environment that are widening the basis of competitive advantage from issues internal to the firm to issues of external relationships among firms. The market is rewarding firms that, having learned how to specialize, then learn how to link their capabilities with other specializing firms. Especially at the high-value ends of markets, goods and services increasingly depend on groups of companies working together in multiple and dynamic relationships (Harrison 1994).

Applying these lessons to job development and employment programs means building on the relationships among firms as a central element of the strategy. Helping complementary and interdependent firms work together to devise human resource strategies that build on their interdependencies offers access to the internal labor market of firms that conventional employment and training policy has been denied.

Firms themselves are groping toward these collective approaches. On one hand, they are driven by global competition, flexible technologies, and segmented demand to disaggregate, decentralize, downsize, and specialize in pursuit of agility. But on the other hand, many of those same firms are now finding they must recombine in new ways to pursue joint solutions to common problems in developing human resources. In other words, although becoming leaner and more specialized is a popular path toward success in high-value markets, firms on this path are encountering serious problems in trying to work on their own to develop a skilled workforce.

Firms trying to upgrade the skills of their incumbent workers encounter problems of scale and scope. Working on their own, training costs more money and takes more time than many are prepared to devote. Sometimes firms are also concerned that if they do invest in training their own workers, there may be little to prevent other nearby companies from luring away these better-trained workers with the promise of slightly higher salaries. Getting all the region's firms in a sector to work together helps to provide economies of scale and scope and begins to get around the problem of such "free riders." Interfirm cooperation is a critical strategy for upgrading the skills of current workers, attracting qualified new job entrants to positions with very high skill requirements, and helping the poor and poorly prepared find family-supporting jobs and move toward economic independence.

The Third Level of Analysis: Business Clusters

A focus on groups of similar firms in major industry groups and industry groups as outlined above supports strategizing about occupational bundles that are of common interest to many roughly similar firms. However, when it comes to the level of analysis that will support even more consequential long-term economic development interventions, it is feasible and necessary for the researcher to define the relevant group as a set of interrelated industries that cut across traditional sector boundaries.

Economists have used a variety of terms to designate critical masses of geographically bounded, related businesses, such as *agglomerations, industrial districts,* and *technology regions.* Here, I use the term *cluster* to define a geographically bounded concentration of similar, related, and complementary businesses, with active channels for business transactions, communications, and dialogue, that share specialized infrastructure, labor markets, and services, and that are faced with common opportunities and threats. Almost always, a cluster includes both suppliers and customers.

For regional economic development, it matters very much when the firms in these multi-industry relationships are geographically concentrated. The tendency of businesses to locate near their suppliers, customers, important services, and competitors—that is, to cluster—occurs in all places and in all industries. Historically, it is a critical source of economic advantage. Specialization within the cluster can promote innovation. Dense concentration can reduce the costs of transactions, stimulate supportive services on an economical basis, and enhance the market visibility of all firms in the region. Information can flow more quickly among firms that enjoy these close relationships than among those that are more isolated from each other. When several sectors closely related to each other are tightly clustered in a region, economic advantage can result. These gains can be derived from lower costs in their multiple transactions with each other that result from proximity and volume. Additionally, especially when economic connections are reinforced by social infrastructure, innovation and knowledge can spread more rapidly through these dense clusters than would be the case if the firms were more scattered and had more limited contact (Porter 1990).

There are two reasons it might be advantageous for job development programs to look beyond sectors to clusters. First, dealing with groupings of interdependent and complementary firms in clusters casts a much wider net of knowledge about market changes, emerging business strategies, and imminent technological advances that might significantly alter labor force requirements over time. Similar firms that employ people in very similar occupations may tend to see their workforce issues through a similar lens. Although they may talk the same language, they may also share the same biases and be limited to the same set of information. Working with clusters of firms who come from multiple sectors is likely to expand the sensing capacity of the group.

For example, working with a group of similar home health care providers who face very similar markets and other competitive pressures might be very satisfying in the sense that they have a shared vocabulary and virtually identical occupations and skill needs, which can facilitate the rapid and accurate exchange of information. Sharing very similar perspectives on the future of their industry, they may be able to move relatively quickly to shared agreement on workforce development strategies. But they may all be similarly unaware of looming technological changes or demand trends that eventually will drive major change in their staff deployment and in the skills they require of entry-level workers. If that group of home health care providers were enlarged to include manufacturers of medical devices, the conversation might widen to how new home care products may change the function of the health care professionals. If the group were expanded to include representatives of health insurance companies, a further range of discussion might be feasible about the probable impacts of managed care on the scope and intensity of home health aide services.

In other words, working with groups of interdependent firms representing a variety of sectors may permit the more careful identification of opportunities for collective workforce development undertakings that will affect the overall competitiveness of the group. Frequently, it will be the transfer of knowledge across traditional and narrow sector lines that stimulates the group to action. It may be at the intersection of product and service groupings that opportunities for constructive cooperation are most apparent.

A second justification for cluster orientation rather than a more narrow focus around similar firms has to do with the way that primary labor markets function. Very frequently, companies hire from their suppliers, vendors, customers, and collaborators. Occupational ladders and employment paths often cross industry group boundaries within clusters. This intersectoral job movement might be especially important for programs looking to place individuals with limited job records and skills into low-wage, entry-level jobs. These starter jobs may not pay wages that will support a family, and therefore the ability to move upward into higher-paying jobs in related firms (i.e., within the cluster) may be a critical element of a successful jobs strategy. A program that is involved in supporting the collective improvement efforts of a cluster of interdependent firms might have a leg up in building multifirm career ladders.

Identifying Clusters

As I have previously observed, firms do not take root and develop at random; they do not grow in isolation from each other. They develop as part of a production system of interrelated and mutually supporting firms and industries. Each class of industry is related though purchases and sales to particular sets of other industries whose needs and capabilities are complementary. In identifying clusters, a researcher would focus on these market relationships. For example, cluster identification can begin with the manipulation of secondary data sources through input-output analysis.[9] This can give a running start on identifying how different groups of industries are linked in a region (DRI/McGraw-Hill 1995). But primary research—talking to the representatives of the firms in the region—

is essential for the researcher to develop a feel for these market relationships, gain a working knowledge of the industries, and widen contacts with key leaders.

Using Cluster Analysis to Shape Strategy

The objective of cluster analysis is to investigate how public-private interventions might strengthen the performance of a cluster in a way that will contribute to enhancing the competitiveness of the firms in the group, stimulate job growth, and increase the likelihood the firms will hire those people that job development agencies wish they would. Unfortunately, cluster development strategies are not as well advanced in the field of economic development as are strategies of sector development. In view of this, it is especially important for job development organizations to build regional coalitions of economic and workforce development organizations that can begin to examine together how they can strengthen their clusters. The needs of firms are multidimensional and interconnected; meeting these needs frequently requires simultaneous change across several fronts, and that depends on concerted action among several development-related organizations.

There are several criteria by which the performance of a cluster might be measured, and by which the feasibility of specific interventions might be assessed. Some of these criteria are outlined briefly in Table 5.2, together with a starter set of considerations that might be addressed to several of the firms in the cluster and that should begin to help to identify opportunities for collaborative undertakings. Sustained involvement with industry leaders in determining how to strengthen, shore up, or broaden access to the various assets or capabilities implied in these dimensions can affect the ability of the firms to compete at progressively higher levels of value added and to grow new job opportunities. Programs that are involved with firms in these dimensions can begin to influence the internal labor markets of these firms in ways that can contribute to the success of the private-public partnership in placing poor people in good jobs.

In Summary

Job development programs that focus exclusively on the supply side of the labor market require little understanding of regional economies. However, any strategy that seeks to use employer demand to "pull" skill development, and that is

TABLE 5.2 Criteria for Measuring Cluster Performance

Category	Questions
Capacity/development of common human resource pool	Do regional education/training institutions offer specialized training to support occupational groups important to cluster? What is the prognosis for the supply of critical, subbaccalaureate skilled worker positions within cluster? Do cluster firms routinely provide upgrade training for employees in these subbaccalaureate positions? Is there a high level of labor mobility in the cluster? What are the occupational connections among firms in the cluster? What aspects of human resource development and job training are of greatest interest to firms across the cluster? Are the firms interested in investigating the normalization of multifirm career ladders?
Ongoing development, rapid diffusion, and widespread adoption of undergirding technological innovation	What are the five most important R&D questions facing the industry? Where is the R&D carried out? How many firms in the cluster have significant R&D budgets? How are research findings disseminated to the industry at large? What is the process by which new ideas and innovations gain commercial acceptance and find specific applications? Do universities/regional institutions have the research capacity to support the needs of these industries? How can research findings be most effectively integrated? What are the five most important recent technological innovations in this industry or set of industries? What percentage of firms in the group have and are proficient in the use of these technologies? What is the focal point through which information about technology is distributed to firms?

(continued)

TABLE 5.2 *Continued*

Category	Questions
Strength of final demand for the producer of consumer goods and services that are the output of the cluster	Does the cluster contain those firms that are most important in understanding and satisfying final market demand? Who or what most influences final demand? What are factors that will most seriously build or erode final demand (i.e., market conditions or technology)? What is the collective capacity of firms in the cluster for influencing the strength of final demand? How have the firms been willing to work together to influence final demand?
Quality of customer-supplier-vendor relationships and their local availability and proximity	Are customer-supplier relationships dominated by just a very few large original equipment manufacturers? What are the characteristics of the relationship between suppliers and original equipment manufacturers? Are suppliers organized across sector boundaries? Do they know each other and work together? What are most the important missing links in the regional supply chain? Can new firms be recruited to serve specific supply niches?
Availability, accessibility, and adequacy of specialized capital and knowledgeable financiers	How well do banks and other lending institutions in the region understand the specialized needs of this cluster? Do the lending institutions meet the cluster's needs for asset financing and working capital? Are there parts of the cluster that are underserved? Do public equity markets support growth within the cluster? Is there private equity financing such as venture capital to support entrepreneurial growth?
Access to specialized business services having intimate familiarity with the markets and production systems of the cluster	Do professions serving the cluster have specialized knowledge of technology, market, and business strategy issues? Are there designers, marketers, and distributors located in close proximity? Are there tight linkages among them? Who does market research and analysis? Are there export development services?

TABLE 5.2 *Continued*

Category	Questions
Global connections	What percentage of goods are exported? How many of the firms export? Are intermediate goods and services exported? Does the cluster participate in international trade shows? Are specialized language services available?
Formal interfirm organization, networks, and alliances	Do regional industry associations include broad cross-sector representation? What services do regional associations provide members? Who or what gets the firms together? Do networks of firms emerge spontaneously in response to market opportunities? Do the firms undertake any joint marketing programs?
Entrepreneurial energy and diversity	How many new business starts have there been annually? Do local firms encourage spin-offs and spin-outs? How does the local business climate support new business formation? Does the cluster include firms that participate in several different markets? Does the scope of participation suggest that the firms collectively have the ability to sense important changes in technology or demand that could quickly affect their competitive positions? Does the diversity of participation involve different underlying technologies?
Shared vision and strong leadership	Who speaks for the cluster? Do key people use similar vocabulary when describing business opportunities and challenges?

therefore concerned with influencing the quantity and quality of that demand, requires a deep understanding of the regional economy. Developing that understanding is an ongoing process of analyzing secondary data and talking to owners, managers, and workers in the firms. This process can be developed at three levels: identifying those individual firms whose contribution to the regional economy is especially important; mobilizing collective action within those particular industry groups that find advantage within the region; and probing the performance of intersectoral business clusters that drive vitality within the regional economy.

Notes

1. If the population of the MSA is larger than 1 million, and if there are distinct "subgroups" of population, it may be subdivided into two or more primary metropolitan statistical areas (PMSAs). The Philadelphia MSA has four PMSAs; Seattle, Milwaukee, and Denver each have two. St. Louis and New Orleans are not subdivided into PMSAs. For some business clusters, researchers might find it very helpful to concentrate on only one of the PMSAs rather than doing all analysis at the larger level. There are occasions when it is not appropriate to look at the whole MSA in considering project design. For example, the commuting capabilities of youth are quite different from the commuting patterns of adults, so for youth development projects it might make sense to deal only with employers in a relatively tightly defined geography (e.g., just the core county).

2. The Standard Industrial Classification system of the U.S. Bureau of Census was last published in manual form in 1972 and last supplemented in 1987. All employers are assigned industry classifications by the Social Security Administration when they apply for Employer Identification Numbers. The manual codes all economic activity into a hierarchy according to similarities among products and services and the technologies and other methods used to produce them. Businesses are coded at the establishment level. An establishment is defined as a location or specific plant, thus a single large firm may have scores of establishments. Each business establishment is assigned a code according to what it mainly produces. The coding is hierarchical, proceeding from 10 large groupings commonly called *sectors* or *industry divisions* to very specific products identified down to a very fine eight-digit level of detail. There are 83 classes of firms at the two-digit level, which are commonly called *major industry groups*. The three-digit levels are referred to as *industry groups*. At the four-digit level, the classifications specify *product lines,* and with each additional digit the product classifications become more fine-grained.

3. There are a few problems with the information in *County Business Patterns*: (a) It is dated, (b) it is not always reliable, and (c) it is not comprehensive (detailed information is sometimes suppressed). In some cases, a partial alternative to the limitations of this publication is available in the form of ES-202 reports at a state's office of employment services. All employers are required to submit quarterly reports for unemployment compensation purposes with their state employment service office. These reports provide information at the four-digit SIC level about employment and wages with no deletion or suppression, and they are more up-to-date than the *County Business Pattern* reports. A researcher will most certainly be able to get 2-year-old information, and perhaps even 1-year-old information. Most state employment offices began to record this information in easily accessible formats, first on tapes and then on floppy disks, in the early or mid-1980s. By now, a researcher should be able to get a 10-year series. The ES-202 reports are not summarized into public reports and are not considered public information. However, most state employment offices will grant access to individuals who are conducting research that has a public policy objective or consequence. The obvious advantages of using the ES-202 files are greater accuracy and more current information that may suggest unfolding trends in these sectors. But even if a researcher is

stuck with slightly out-of-date information at this stage of analysis, the differences are not likely to be dramatic for the purposes of this summary analysis.

4. Employment data are collected by the U.S. Census Bureau and from the U.S. Bureau of Labor Statistics. The main data sources are the decennial censuses, the Current Population Survey, and the Current Employment Statistics Program. The first of these contains the most complete data, but is only available for 1990, 1980, 1970, and so on. The latter two surveys are conducted monthly on samples of individuals and establishments, respectively. Employment and unemployment data from these two surveys are reported monthly in the BLS periodical *Employment and Earnings,* which should be available at a good local library. Data are reported for each state and for metropolitan areas within each state. Two other BLS publications may be of interest: *Monthly Labor Review* provides employment and inflation data in less detailed form than *Employment and Earnings,* and the *Handbook of Methods* provides detailed information on BLS statistics series. The Bureau of Labor Statistics produces and publishes several periodic reports about employment, earnings, prices, productivity, and technology by major industry sectors on a national basis, and for states and MSAs. It also issues long-term employment projections for more than 200 industry sectors in which it forecasts demand and utilization. Among several other publications that might be helpful to researchers, both for analyzing key trends in certain sectors and clusters and for profiling labor market issues, are the following: *Area Wage Surveys* by MSAs (annual); *Geographic Profiles of Employment and Unemployment* for the larger MSAs and some central cities (annual); *Industry Wage Surveys* for the nation as a whole and for most MSAs (annual); and *Occupational Outlook Quarterly.*

5. There are several other publications available that can be helpful and that offer far more comprehensive guides to regional economic analysis. One of the best is a detailed manual by McLean and Voyteck (1992).

6. Location quotient analysis is a device for gauging the relative specialization of the region in selected industry categories. If employment is to be used as the relevant variable, one calculates the location quotient by measuring the percentage of a region's total of employment that is found within a particular industry compared with (divided by) the same ration for the nation as a whole or the comparison region. If the resulting ration is greater than one, it can be assumed that industry is a net exporter of its production to other regions; the region is relatively specialized in that industry. The higher that number, the greater the location quotient and the more significant the regional specialization. For example, about 10% of Milwaukee's jobs are in the machinery and metal products industry groups, compared with only 3% for the United States as a whole. This tells us that, in terms of jobs, Milwaukee is more than three times more specialized in these activities than we would have expected if all we knew were the national averages.

Of course, a region may have a very high location quotient in an industry group that provides little employment. For example, St. Louis has a very high location quotient for "water transportation of freight" (not too astonishing for the region where the Missouri River meets the mighty Mississippi). But that three-digit classification employs only 927 souls. Therefore, it is a good idea to select for analysis only those industry groups where the share of employment and the location quotient are both high.

It can be interesting to compare location quotients based on employment with location quotients for the same industry that are based on the number of establishments or on average payroll or even value added per employee. This can indicate, in the first instance, where a particular industry group in the region is more disaggregated than it is in the nation or other referent location. In the second and third instances, this analysis will suggest whether firms in a certain industry group in the region are more productive or simply pay higher wages than firms in other locations.

7. Shift-share analysis is a method of data analysis used to isolate the effect of regional differences on growth from those that affect the industry at the national level. It indicates whether a region's share of employment or number of establishments in a particular industry in the area is changing faster or slower than in the nation as a whole. Obviously, shift-share analysis requires careful selection of appropriate intervals to measure these changes. One should generally perform shift-share analysis (and location quotient analysis) at the level of major industry groups (two-digit SIC level) or, better, industry groups (three-digit level).

8. The Department of Commerce assigns individuals to develop these annual updates, and these persons may be able to track down other published information.

9. Through intensive sample survey research, government agencies have periodically measured the portion of its input that each industry group (at the three-digit SIC level) receives from other industries and the portion of its output that it sells to other industry groups. Input-output tables show these historic relationships among various industries. Unfortunately, the process by which the tables are developed is very complex, and it is unlikely that such tables would be available for specific MSAs. A few important words of caution are in order here. The calculations that produce the input-output tables are not nearly as precise as the numbers might lead one to believe. Plus, they are several years out of date; technological change and market shifts have altered the production equation dramatically for many industries. Moreover, the national tables are really just averages of all the different regions. Although the national tables may show that aerospace manufacturing generally draws heavily on the production of machine tools, careful regional analysis in, for example, Seattle and St. Louis (both with aerospace concentrations) might reveal far more machine tool companies located in one than in the other. Finally, and most important, these tables reveal only buy-sell relationships among firms. As has been previously noted, buying and selling data do not capture some of the most intimate and important interdependent connections among firms in a particular cluster.

References

Clark, P. and S. L. Dawson, with A. J. Kays, F. Molina, and R. Surpin. 1995. *Jobs and the urban poor: Privately initiated sectoral strategies.* Washington, DC: Aspen Institute.

DRI/McGraw-Hill. 1995. *America's clusters: Background paper for June 1995 conference in Sedona, Arizona.* New York: DRI/McGraw-Hill.

Harrison, B. 1994. *Lean and mean: The changing landscape of corporate power in the age of flexibility.* New York: Basic Books.

Harrison, B., M. Weiss, and J. Gant. 1995. *Building bridges: CDCs and the world of employment training.* New York: Ford Foundation.

McLean, M. L., and K. P. Voyteck. 1992. *Understanding your economy: Using analysis to guide local strategic planning.* Chicago: American Planning Association.

Osterman, P. 1988. *Employment futures: Reorganization, dislocation, and public policy.* New York: Oxford University Press.

Porter, M. 1990. *The competitive advantage of nations.* New York: Free Press.

Rosenfeld, S. A. 1995a. *Industrial strength strategies: Business clusters and public policy.* Washington, DC: Aspen Institute.

Rosenfeld, S. A. 1995b. *New technologies and new skills: Two year colleges at the vanguard of modernization.* Chapel Hill, NC: Regional Technology Strategies.

6

Labor Market Profiling

Case Studies of Information-Gathering Techniques for Employment Projects

PEGGY CLARK
AMY J. KAYS

This chapter focuses on labor market profiling and participant assessment. Labor market profiling, as the term is used by employment programs, is the process of gathering information on industries, employers, occupations, and participants. Profiling enables an employment practitioner to serve more strategically as a bridge connecting the labor market to the target labor force. The most successful employment projects are those that are highly customized for employers and job seekers (U.S. Department of Labor 1995). For example, welfare populations include both job-ready and hard-to-employ groups who face multiple barriers. In this chapter we document a range of secondary and primary data collection methods for the design and implementation stages of employment projects that have proven useful to practitioners.

In order to explore labor market profiling, we first selected tools and techniques developed by researchers and community-based groups that have proven

NOTE: This chapter is excerpted, by permission, from Peggy Clark and Amy J. Kays, *Labor Market Profiling: Case Studies of Innovative Information-Gathering Techniques for Employment Projects,* copyright 1997 The Aspen Institute, Washington, D.C.

effective for profiling the labor market. Subsequently, we created an organizing framework to provide a context and structure for the various tools, techniques, and approaches described in this chapter. The framework identifies the stages in the life cycle of an employment project. The first is the *design stage,* when information needs are great and the program has not yet selected specific industries, occupations, or target groups, or identified the set of barriers the project will attempt to overcome. The second is the *implementation stage,* when the program design is complete and the program is operating and actively interacting with participants over time. The framework assumes that data will be collected on both the labor force—the targeted community participants—and on the labor market, including industries, employers, and the specific occupations within these industries.

The framework describes both *primary data collection techniques,* which involve the gathering of information directly from program participants, employers, and others; and *secondary data collection techniques,* which involve surveys, interviews, and data collection completed by other researchers. Both of these techniques are valuable. Secondary sources capture information on large numbers of people, and can provide detailed information on a large geographic area or clusters of types of people. Established secondary sources, such as the decennial U.S. Census and state unemployment files, are highly credible and serve as leading sources of reliable data for researchers and practitioners. Primary techniques, on the other hand, are generally more targeted in terms of the scope of the subject areas and the number of people. They do, however, afford access to direct and timely information on the characteristics and labor market issues of people who are working with an employment program.

Collecting Data on Jobs: Case Studies and Sources

There are a wide range of sources available, many of which are governmental, from which a researcher can collect information on the labor market in a given community. Most of these sources identify industries, specific firms, and occupations within industries, yet it is difficult to break these data down to particular geographic areas or occupations. Two case studies demonstrate the use of these data sources and other primary collection techniques. In the first case, secondary data were used to identify jobs for a particular target group in a public housing project. In the second, focus groups were used to generate primary data from employers for use in the design of a new employment program.

Case Study: Identifying Jobs for Public Housing Residents

In 1995, the Summit County, Ohio, Department of Human Services asked Case Western Reserve University's Center for Urban Poverty and Social Change to identify, locate, and analyze a set of occupations that could provide suitable job opportunities to residents of Pinewood Estates, a public housing project in the Cleveland-Akron metropolitan area (Leete and Bania 1995). This research project assessed the occupations in the Cleveland-Akron area and the skills of the potential labor supply of residents of Twinsburg, where Pinewood Estates is located. Researchers used the ES-202 establishment file from the state employment security office to determine the current locations of all employers by zip codes and industrial classifications. They estimated the amount of employment by occupation using the average occupational staffing pattern for firms in a given industry occupation. They then allocated projected numbers of job openings by occupation for selected zip codes in the Cleveland-Akron labor market using projections from the Ohio Bureau of Employment Services and the U.S. Department of Labor's Bureau of Labor Statistics data. The result was a detailed estimate of the expected number of job openings per year for more than 400 occupational categories and more than 500 zip codes in the metropolitan area.[1]

The next task was to determine the specific skills requirements for these projected jobs. Recognizing that they would need to link these findings back to the population of Pinewood Estates, the researchers categorized occupations according to the amount of education and training needed for each job. They used the following categories:

- *Entry-level occupations,* which require 11 to 12 years of formal schooling and 6 months or less of training
- *Short-term training occupations,* which require 12 to 13 years of formal education and/or 6-12 months of training
- *Long-term training occupations,* which require 13 to 14 years of formal education and/or 1-2 years of training
- *Very-high-skill occupations,* which include all other occupations in the labor market

The researchers then profiled their target population. They obtained information on the demographic and educational characteristics of the client population using data from the Department of Human Services of Summit County on Pinewood Estates residents who had accessed department services. The researchers concluded that 60% of Pinewood Estates residents could fill only entry-level occupations or below. Yet they found that 71% of these entry-level occupations paid very low wages—not enough to bring a family above the poverty line. Simply providing job search assistance would not alleviate poverty for the majority of Pinewood Estates residents. The education and skill gap was simply too great to overcome with minimal assistance. The client population needed 2 to 3 years of training and skill upgrading to achieve long-term economic security.

Case Study: Using Focus Groups to Design a New Employment Program

In 1993, Warren/Conner Community Development Corporation, an established community development corporation in Detroit, launched a new employment initiative, the Partnership for Economic Independence (PEI). As part of its planning process, PEI engaged a consultant to conduct focus groups with Detroit employers and local employment and training program representatives to inform the design of the new program (Hansen 1993a, 1993b).

Focus groups with representatives of Detroit businesses explored the perceptions of employers about the local workforce and their current hiring practices. Participants in the focus groups were asked questions about recruitment and hiring practices, training, worker attitudes and expectations, substance abuse, employee social and personal problems, and neighborhoods. The research revealed that the employers rarely conduct formal recruitment for entry-level workers. Most have stable workforces, with little turnover. The employers also stated that a person coming in from the street is not likely to find employment in their companies; obtaining employment generally requires a referral from someone already employed with the company. Finally, most said they would like to hire from the neighborhood but felt that the residents have personal problems that often make them unsuitable employees.

The following excerpt from the focus group of Detroit employers further elaborates:

Moderator: Earlier you lamented the fact that you had few employees from your immediate neighborhood.

Respondent 3: Oh, we get plenty of applicants, but we haven't hired off the street in a long time; we don't really want to take the chance. We make him [the applicant] come back four or five times before we will consider him. He may just want something for tomorrow. He may just have a bill to pay. It may rain tomorrow. So you have to get them to come back three or four times. I had one this morning. He was there yesterday. He walked back from Jefferson and the Boulevard today to tell me that his telephone number had been changed. Well, that's a good sign. He was there yesterday, and he was a referral. And he walked. That's good. He's really looking for a job.

Collecting Data on the Target Population: Case Studies and Sources

Information on the characteristics of local residents is critical for determining the target population of an employment program. A detailed knowledge of the issues confronted by the target labor force—together with an understanding of the labor force requirements of local employers—enables employment practitioners to design and implement programs that can help participants get and keep good jobs. In this section, we describe a range of sources of secondary data and discuss their strengths and limitations for profiling a community's workforce. Following this, we present a case study illustrating how primary data collection (interviews and focus groups with local residents) can supplement the broad picture provided by secondary data.

At the design stage of an employment project, various sources of secondary data can provide a general picture of the residents living in a defined geographic area. These sources help identify patterns in labor force characteristics and potential barriers to employment. A few critical indicators determine the broad workforce characteristics of a target population. These include education levels, employment status, and work experience. Additional important indicators not generally considered to be related to employment include gender, marital status, number and age of children, language spoken at home, and race. These other indicators may suggest a need for additional program services, such as child-care or language services.

The main source of secondary data describing the demographic characteristics of a target population is the U.S. Census.[2] The census is conducted every 10 years, thus the most recent data were collected in 1990. Census data are available in printed reports, on CD-ROM in major libraries, on computer tape files on mainframe computers, and from the U.S. Census Bureau's state data centers and regional offices, where census specialists prepare custom data products.

The Census Bureau collects data on age, sex, race, Hispanic origin, and household relationship for 100% of the population. More detailed information, including years of education, marital status, occupation, income, and distance of commute to work, is collected from a sample designed to represent the nation as well as smaller geographic units, such as cities, neighborhoods, and rural counties.[3]

Census data are organized in a geographical hierarchy, from summary data on the entire U.S. population down to blocks that have an average population of 85 residents. The most readily available data for location-specific research are published in printed reports for metropolitan statistical areas (MSAs), which are geographic areas that include both cities and their surrounding metropolitan counties. Although it is easiest to find and interpret data at the MSA level, most community-based employment projects target smaller areas, such as neighborhoods or groups of neighborhoods. For this finer level of exploration, researchers must determine which census tracts cover the targeted area and obtain data at the census tract level.[4]

The most detailed data describing demographic and economic characteristics of the target population are those collected by the Census Bureau for a sample of the population. These data are available from the Census Bureau on Summary Tape Files 3 and 4 (STF3 and STF4). Because tape files are difficult to access (requiring mainframe computers and special programming skills), an expedient way of getting data at this level of detail is to request a custom data product or user-defined area tabulation directly from a census specialist at a state data center or regional U.S. Census office. Data at the MSA level and, for some locations, at the census tract level, are also available on CD-ROM. This census tool is called PUMS (Public Use Microdata Sample) and is a 1% sample of all census data for an MSA.

Case Study: Identifying Job Opportunities for the Hard to Employ in Chicago

In their work with job training and placement agencies in Chicago, Nikolas Theodore and Virginia Carlson found in the mid-1990s that many local agencies were struggling to match job seekers with employment opportunities in the context of a radically changed local labor market (Theodore and Carlson 1996). Information that was once available on leading industries, employers, and job requirements was no longer sufficient. In an attempt to address this need, Theodore and Carlson developed a comprehensive approach to gathering labor market information to inform local practice. Although their work focused on techniques for gathering information on both employers and community residents, we describe here only their profiling of a target population.

They began by determining the characteristics of area residents. Their principal source was the 1990 U.S. Census, from which they obtained data on two critical indicators: education levels and occupations. Using Summary Tape Files 3 and 4, they identified the occupations and industries of residents by census tract. STF3 reports summary data, and STF4 reports industry and occupation by race and gender. The second set of census data they used came from the Public Use Microdata Sample. PUMS data are individual records reported for a representative sample of people living in areas that are aggregates of census tract areas; these aggregate areas are called PUMS areas. PUMS data provide more detailed occupational and geographic detail than do Summary Tape Files data.

In order to match the resident target population to occupations, Theodore and Carlson categorized available occupations according to the following characteristics: skills requirements, average wages, projections of employment change, and gender breakdown. They believed that the most important information on any given occupation is the level of education and training required.

Theodore and Carlson used these techniques to provide assistance to many employment and training agencies in the design stages of their programs. An example is the Spanish Coalition for Jobs in Chicago, which wanted to identify well-paying occupations for the Mexican immigrant community it serves. Through an analysis of the client population, the Spanish Coalition for Jobs identified the occupations of electronic and electrical technicians as appropriate for participants who were interested in further skill training. The coalition requested that Theodore and Carlson help to examine local industries that had these jobs and to determine the specific occupational requirements of these employers.

Case Study: Using a Community-Based Organizing Network to Gather Data on the Labor Force

Project QUEST (Quality Employment through Skills Training) is an employment and training program that assists low-income residents of San Antonio, Texas, in completing long-term training for skilled occupations (see Lautsch and Osterman, Chapter 12, this volume).[5] QUEST's innovative design is the result of hundreds of interviews and meetings with local residents and employers.[6] Organizers spent 2 years researching why the economy no longer worked for low-skilled workers and shared their findings with local residents. Simultaneously, they listened to the concerns San Antonians voiced about their ability to support their families with available jobs and their experiences with the public employment programs that were intended to assist them in finding work. What they heard was that the institutions set up to help these individuals had often left them with broken promises, student loan debt, and poor employment prospects.

From these "house meetings" with residents in their neighborhoods, interviews with employers, and information from studies and statistics, community leaders concluded that many local employers needed good workers and that local people were willing to make the commitments and sacrifices that would be required to complete the long-term training needed to get good jobs.

> The house meetings uncovered reasons people dropped out of traditional training programs, which in turn suggested further components of QUEST. Trainees often felt that no one cared about their progress, and many people had to drop out of classes because of family problems. "We needed counselors that were also caseworkers. We needed to have remedial schooling because some of the people had been out of school for 15 years, and they had forgotten their math. Women with small children needed day care during classes." These components were added. (COPS and Metro Alliance 1994, 10)

The direct comments of community residents as well as extensive investigations of local labor market employers, Job Training Partnership Act contract training facilities, and other education providers (such as vocational-technical schools and community colleges) shaped the design of Project QUEST. Through this intensive community-based primary research, QUEST came to understand *why* residents could not get the training and support they needed to succeed in higher-skilled occupations. This research informed not only Project QUEST's founding principles and design but also provided the impetus for community organizing to obtain adequate training resources.

Defining the Target Population

Once the broad characteristics of the labor force in a geographic target area are identified, program designs must define more specifically the parameters of the target population. This effort is essential; it specifies who will and will not be served by a program. This task is particularly sensitive for community-based employment programs that are trying to meet the needs of disadvantaged people and of employers who want to hire the most job ready among the target population. Defining a target population requires reflecting on the mission, goals, and values of the sponsoring agency; articulating a sense of the group of people who would be served; and weighing this against the requirements of the labor market.

In reality, participant selection is rarely this simple or clear. However, leading employment practitioners agree that some parameters are needed to define those community residents who can and cannot be served by a program given its resources. The boundaries that need to be established in order to define a target group will generally include some or all of the following key defining characteristics: education level, math and reading testing level, proficiency in English, employment history, and other barriers to employment that may make it difficult for an individual to succeed in employment, such as lack of child care.

The employment and training programs highlighted in case studies in this chapter demonstrate a wide diversity of program entrance requirements that define their target populations. Project QUEST is the only one that requires a high school diploma and a demonstrated commitment based on an interview with a community organizer. Cooperative Home Care Associates in the South Bronx has no formal educational requirement, but participants must demonstrate a caring attitude, responsibility, willingness to learn, and interest in cooperative membership. Neither of the other two programs—Partnership for Economic Independence in Detroit and Project Match in Chicago—has formal requirements; both serve the very disadvantaged, and most participants have extensive labor market barriers.

Information Gathering in the Implementation Stage of an Employment Project: Case Studies and Sources

At implementation, employment projects need tools and techniques to assess participants and to monitor their progress throughout training activities and job placement. A broad array of assessment tools are available. Standardized paper-and-pencil tests measure basic academic skills and knowledge. Packages designed for specific occupations teach and measure progress toward the achievement of technical competence. Qualitative methods identify and improve behavior and attitudes necessary for success in the workplace.

There are many standardized tests intended to determine individuals' skill levels, educational achievement levels, and standing in other areas. There is, however, considerable controversy about the usefulness and validity of many standardized tests of competencies for employment purposes. In a review of assessment tools, the Urban Institute pointed out in 1990 that there is still ongoing debate and concern about how well existing tests predict job entry, job performance, and performance in occupational training (Yudd and Nightingale 1990).

Some community-based programs have found that pencil-and-paper tests are difficult for people who have not performed well in formal classroom settings, or who have language or other communication barriers. Such tests also may not be effective for measuring a person's potential for achievement. These programs have developed their own tools designed to measure some very specific aptitudes or attitudes. The following case study of Cooperative Home Care Associates describes the tools and techniques developed in this program to assess an applicant's chances of success (see also Dawson, Chapter 7, this volume). Another case study, Project Match in Chicago, highlights how a welfare-to-work program has designed an interactive participant monitoring tool that is useful to both the interviewer and the participant.

Case Study: A Participant Assessment Interview

Cooperative Home Care Associates (CHCA) in the South Bronx is a leading worker-owned employment program that trains low-income community members for jobs as home health care aides. CHCA uses an intensive one-on-one interview as an assessment tool. The CHCA staff interviewer guides the applicant through a series of situational and hypothetical questions in order to determine the applicant's aptitude for a job as a home care aide. This process utilizes the CHCA Interview Assessment Sheet (see Figure 6.1).

The Assessment Sheet is composed of five sections: attitude regarding patient care, job skills, work history, personal traits, and attitude about the worker-ownership structure of CHCA. Within these five areas, specific situational questions are posed and sample positive responses are given. The applicant is rated according to his or her answers in the five areas, and the interviewer assesses whether or not the applicant has the right mix of aptitudes and skills to be a successful CHCA home care aide. An illustrative excerpt from the CHCA Assessment Interview Guide appears below.

Labor Market Profiling 115

APPLICANT'S NAME _____ DATE _____

QUALIFICATIONS	VALUE	RATING	COMMENTS
ATTITUDE RE: PATIENT CARE • Likes/dislikes about home care work • Working with different cultural/ethnic groups • Handling difficult patients • Interest in patient beyond scope of required services • Fear of certain diseases or other prejudices (e.g., neighborhoods, pets, etc.)	35		
JOB SKILLS: Demonstrates ability in: • Organizing workday/doing tasks • Analyzing/resolving problems • Handling difficult patients • Handling different diets • Changing dressings, taking TPR, ROM, safety measures, etc. • Different diseases, i.e., diabetes, hypertension, etc. • Observant re: health/home needs	20		
WORK HISTORY including: • Employment gaps • Differences in types of employment • Relevance of work history • Reasons for leaving jobs • Preference for hours • Relevance of unpaid work	15		
PERSONAL TRAITS • Appearance • Motivation/enthusiasm/interest • Punctuality • Passive/aggressive manner/voice tone • Poor/good listener • Patient/impatient	20		
ATTITUDE ABOUT COMPANY • Worker-ownership structure • Team meetings • Willingness to participate as member of cooperative	10		
TOTAL	100		

Figure 6.1. Cooperative Home Care Associates Trainee Interview Assessment

116   TECHNIQUES

- **Attitude Regarding Home Care Work**

Have you ever cared for anyone who was very ill before? Who? How long? What was their illness? What did you do while caring for them?

>(If the person has cared for someone over a long-term period, use that experience as the context for posing some of the situational questions, for example, were there times when the person refused a bath or grooming?)

What do you get out of doing this type of work? What do you think will be some of the harder parts of this job? Why?

>*Responses:*
>
>Discusses experiences in a caring manner (verbally and nonverbally).
>
>Shows awareness about specific health problems.
>
>Helped with personal care is a plus! Shows awareness and responsiveness to mood changes due to illness.

Situational question:

>You walk into the client's home on your first day on this case. The house is extremely dirty. The client also looks like she has not been taking care of her personal grooming. After you introduce yourself, she tells you that her sister will be coming by to clean her on a regular basis and she doesn't want you to worry about that part of your job. All she wants you to do is clean the house and do her laundry and shopping. What would you do?
>
>*Responses:*
>
>Discusses duties calmly with client.
>
>Tries to persuade her in a caring manner to allow home health aide to do her full job.
>
>Ultimately, calls her coordinator to report problem if unresolved.

Case Study: Designing Innovative Assessment and Monitoring Tools to Empower Program Participants

Project Match is a leading welfare-to-work program that serves residents of the Cabrini-Green public housing project in Chicago.[7] Working with very-hard-to-employ long-term welfare recipients, Project Match has pioneered new approaches to participant assessment and monitoring. Project Match's director, Toby Herr, designed assessment and monitoring tools to help participants make progress toward moving out of poverty. Recognizing that this path is marked by small, incremental steps, rather than one large transformation, Herr created tools that allow participants to track their individual routes of incremental gains.

Herr discusses her philosophy regarding assessment techniques used in employment projects:

> In the context of welfare-to-work policy and programming, the word "assessment" encompasses questions related to sequencing (should a particular parent begin with school, work, or job readiness?); what and how much information will be used to make the decision (standardized tests? a career inventory?); and who will make the decision (a caseworker alone? the parent and caseworker together?). At Project Match, unlike most other welfare-to-work programs, we do not attach much importance to an initial career assessment. We know that easily measured characteristics, such as basic skill levels and years of education, are not reliable predictors of readiness for work. In addition, it is only after a counselor has had the opportunity to work with and learn about participants over time that any meaningful assessment can occur. (in Herr and Halpern with Majeske 1993)

Project Match has developed a "visual metaphor" called the Incremental Ladder to Economic Independence (Herr and Halpern with Conrad 1991, 3-4). The ladder is used in one-on-one discussions between Project Match staff members and participants to help participants visualize the specific steps they can take, or sets of activities they can participate in, to move toward economic self-sufficiency. The concept of a ladder is based in the belief that, for many long-term welfare recipients, the move from being on welfare and not working to holding down a full-time job without any welfare assistance is a huge step. In fact, research shows that the majority of long-term welfare recipients who *do* get jobs keep them only for a few months at most. The leaders of Project Match recognized that they needed to start where people were, and to give participants regular positive reinforcement and recognition of the small steps they were taking that involved initiative and responsibility toward their children.

Summary and Key Issues

The task of matching disadvantaged people to good jobs has grown ever more complex over time. As the urban labor market has grown smaller for people with limited education and skills, community-based organizations have found that there is a greater need for good jobs for the people in the communities they serve, as well as a greater challenge in finding such jobs. One critical part of this challenge is the need for detailed and current information on the labor market and the labor force. In this chapter we have highlighted some approaches that academic researchers and community-based practitioners have pioneered. Four lessons emerge.

First, *programs need timely information.* Traditional industry sector analyses do not generally provide timely local information. Although these secondary sources provide a good foundation for understanding the general economic base of an economy, businesses and hiring practices are extremely dynamic and respond to a variety of external factors. More strategic primary data, such as interviews with employers, are critical to getting a complete picture.

Second, *specific occupations require a mix of worker aptitudes, skills, and interests.* For this reason, information on basic indicators such as education level attained is often inadequate for assessing how a participant will perform in an employment program. For example, Cooperative Home Care Associates has found through experience that specific personality traits and experience caring for sick people are the best predictors of success as a home care aide.

Third, *creativity in the design of primary data-gathering techniques can be used to help build a strong constituency for an employment project.* For example, Project QUEST's extensive research in San Antonio communities and with leading employers and policy makers offered these interested parties a chance to have a voice in the design of the project. This was instrumental in the establishment of an effective program design as well as in helping these constituencies feel ownership over and enduring interest in the success of the project.

Finally, *highly effective employment projects utilize ongoing information gathering with participants as a matter of course.* Project Match's self-sufficiency ladder is an integral part of the ongoing dialogue between caseworkers and participants. It serves to clarify steps and progress toward articulated goals. Most important, it is useful to both the caseworker and the participant.

Employment projects should develop and use a variety of tools and techniques to gather information on the local labor market and their targeted labor force. We have attempted to present a range of techniques developed by both academic researchers and community-based groups. We hope that this discussion will begin to define labor market profiling as an important area of work that will gain recognition, and that future work will document a broader array of approaches that are useful in helping low-income people find work.

Notes

1. Leete and Bania (1995) used three leading measures to determine occupational requirements: the General Educational Development (GED), the specific vocational preparation (SVP) required for an occupation, and the actual education of those currently employed in those occupations. GED and SVP are measures of job content developed by the Department of Labor, and are listed for 12,000 occupations in the *Dictionary of Occupational Titles*. To determine the actual characteristics of workers employed in given occupations, they used the 1990 Census of Population and Housing. With this information, they then rated the available jobs according to the four categories described on page 107, in order to create a clear picture of how much education and training would be required for residents to access these jobs.

2. The primary guide to the 1990 census is the *1990 Census of Population and Housing Guide. Part A, Text* covers information on data products, geographic terms and products, where to find assistance, population and housing concepts, and census procedures. *Part B, Glossary* defines terms used in the census, describes the questions asked on the census and includes a sample of the questionnaires, and provides technical terms associated with collection, processing, and tabulation, including terms used in working with data on electronic media. *Part C* is an index to the 1990 Census Summary Tape Files. These publications are available from the Superintendent of Documents (fax number [202] 512-2250).

3. Collecting data on rural areas presents unique challenges due to large geographic areas, small populations, and the need to preserve the confidentiality of census respondents. For very helpful information on conducting research in rural areas, see Salant and Waller's *Guide to Rural Data* (1995), which is newly revised and is available from Island Press in Washington, D.C. (phone number [800] 828-1302).

4. Three reports that are useful for this level of research are *Social and Economic Characteristics for Metropolitan Statistical Areas, Social and Economic Characteristics for Urbanized Areas,* and *Population and Housing Characteristics for Census Tracts and Block Numbering Areas.* All are published by the U.S. Bureau of the Census.

5. This summary draws heavily from *Investing in People: The Story of Project QUEST* (COPS and Metro Alliance 1994).

6. Project QUEST was founded on the grassroots organizing efforts of Communities Organized for Public Service (COPS) and Metro Alliance—two Industrial Areas Foundation organizations in San Antonio.

7. This summary draws heavily from three Project Match publications: Herr and Halpern with Conrad (1991), Herr and Wagner with Halpern and Majeske (1995), and Herr and Halpern with Majeske (1993).

References

COPS and Metro Alliance. 1994. *Investing in people: The story of Project QUEST.* San Antonio, TX: Communities Organized for Public Service and Metro Alliance.

Hansen, K. 1993a. Report of Focus Group I: Detroit east side establishments. Unpublished manuscript, August.

Hansen, K. 1993b. Report of Focus Group II: Job training and placement specialists. Unpublished manuscript, October.

Herr T., and R. Halpern, with A. Conrad. 1991. *Changing what counts: Rethinking the journey out of welfare.* Evanston, IL: Center for Urban Affairs and Policy Research, Northwestern University.

Herr T., and R. Halpern, with R. Majeske. 1993. *Bridging the worlds of Head Start and welfare-to-work: Building a two-generation self-sufficiency program from the ground up.* Chicago: Project Match, Erikson Institute.

Herr, T., and S. L. Wagner, with R. Halpern and R. Majeske. 1995. *Understanding case management in a welfare-to-work program: The Project Match experience.* Draft 4. Chicago: Project Match, Erikson Institute.

Leete, L., and N. Bania. 1995. *Assessment of the geographic distribution and skill requirements in the Cleveland-Akron metropolitan area.* Cleveland, OH: Center for Urban Poverty and Social Change, Mandel School of Applied Social Science, Case Western Reserve University.

Salant, P., and A. J. Waller. 1995. *Guide to rural data.* Washington, DC: Island.

Theodore, N. C., and V. L. Carlson. 1996. *Targeting job opportunities for the hard to employ: Developing measures of local employment.* Chicago: Chicago Urban League.

U.S. Department of Labor. 1995. *What's working (and what's not): A summary of research on the economic impacts of employment and training programs.* Washington, DC: U.S. Department of Labor.

Yudd, R., and D. S. Nightingale. 1990. *The availability of information for defining and assessing basic skills required for specific occupations.* Washington, DC: Urban Institute.

PART III

EXAMPLES

7

Start-Ups and Replication

STEVEN L. DAWSON

Originating in 1985 in a gritty South Bronx office—one that looked unnervingly like it belonged to a private detective on the skids—the Cooperative Health Care Network now links three profitable home health care cooperatives in the inner cities of the South Bronx, Philadelphia, and Boston (Nye and Schramm 1994). This federation of employee-owned businesses stands as a rather rare exception to a long and frustrating history of community-based enterprise creation (Lehmann 1994; Vidal 1992). Together, these three cooperatives now employ more than 500 paraprofessional home care aides, nearly all of whom are African American and Latina women. Of these, more than 400 were formerly dependent upon public assistance.

The various actors within our story will be difficult to follow, however, without a scorecard. Figure 7.1 illustrates the relationships among the entities described below. Cooperative Home Care Associates (CHCA) is the initial cooperative, started in the South Bronx in 1985, which now employs 390 home health aides. CHCA and the other two enterprises are structured as worker-owned cooperatives, in which each employee—from the president to every home health aide—has the option of owning one voting share of stock.

Home Care Associates (HCA) is the first replication site, started in Philadelphia in 1993, which now employs 70 aides. Cooperative Home Care of Boston (CHCB) is the second replication site, started in Boston in 1994, which now employs 60 aides.

Figure

```
┌─────────────────────────────────────────────────────────────┐
│   Cooperative        Home Care        Cooperative           │
│   Home Care          Associates       Home Care             │
│   Associates (CHCA)  (HCA)            of Boston (CHCB)      │
│   ~ initial site ~   ~ first          ~ second              │
│   SOUTH BRONX        replication ~    replication ~         │
│                      PHILADELPHIA     BOSTON                │
│                                                             │
│           Cooperative Health Care Network                   │
│                ~ informal federation ~                      │
└─────────────────────────────────────────────────────────────┘
┌─────────────────────────────────────────────────────────────┐
│      Paraprofessional Healthcare Institute (Institute)      │
│              - 501 (c)(3) development corporation           │
└─────────────────────────────────────────────────────────────┘
```

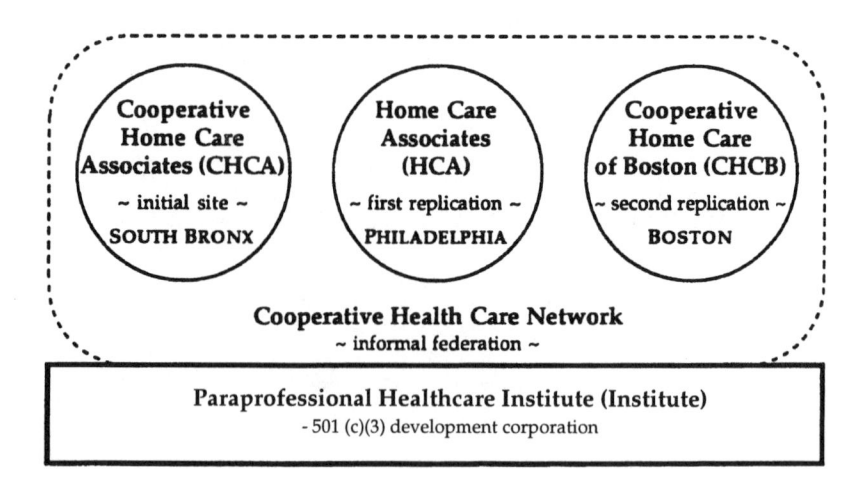

Figure 7.1. The Relationships Among the Elements of the Cooperative Health Care Network

Finally, the Paraprofessional Healthcare Institute (the Institute), of which I serve as president, is the 6-year-old 501(c)(3) nonprofit agency based in the South Bronx that undertook the replication program. The Cooperative Health Care Network is the informal federation that links together the three for-profit cooperatives.

In the remainder of this chapter, I provide a description of the people the Cooperative Health Care Network trains and employs, the types of jobs they perform, and the three cooperatives that employ them. I also provide a brief chronology, lessons learned about the process of enterprise replication, and the key "design elements" of our community/labor-based enterprise model.

Context

Participants

The Cooperative Network's typical entry-level trainee is a woman of color between the ages of 22 and 55. She is single, the mother or guardian of young children, and was dependent on public assistance before participating in the training and employment program. Although she may have done poorly in school—math and reading skills typically range between the fourth- and eighth-grade levels—she is nonetheless an extremely resourceful and caring individual.

Home Care Jobs

Home health care provides one of the few types of jobs available to women who have little formal schooling and limited job experience; in fact, in recent years home care has become one of the first stops off public assistance for literally hundreds of thousands of inner-city, low-income women.

Home health care aides provide paraprofessional care—hands-on assistance with health care needs, bathing, toileting, grooming, and meal preparation for their ill and elderly homebound clients. These and other closely related frontline health care jobs—home attendant and personal care positions, as well as certified nurse's aide positions in hospitals and nursing homes—employ more than 2.2 million people in the United States, of whom 85% are women and 30% are women of color (Himmelstein, Lewontin, and Woolhandler 1996).

Unfortunately, these jobs are of such poor quality that nearly 600,000 medical care workers have family incomes below the federally defined poverty line (Himmelstein, Lewontin, and Woolhandler 1996). Current federal Medicare regulations require no more than 2 weeks of training for these positions, and they make no attempt to ensure that these workers are provided a livable income and adequate benefits—average wages typically range near $6.00 per hour or less (Dow 1991, 1993), and many positions, particularly in the home care industry, are part-time, averaging 25 hours per week (Bayer, Stone, and Friedland 1993). In a bitter irony, 42% of all health care workers in 1993 had no employer contribution toward health insurance at all, up from 37% in 1989 (Himmelstein, Lewontin, and Woolhandler 1996).

In particular, the job of a home care aide is extraordinarily demanding—requiring that she care for her client alone, in the client's home, with almost no on-site supervision. She must also be prepared to resolve a range of unpredictable problems—from calming an irate family member to responding to a sudden health crisis. And most important, she must each day be a warm, caring companion to an elderly man or woman who may be insecure, lonely, or disoriented.

Given that such a difficult job is rewarded with poor wages and benefits, it is not surprising that annual turnover of home care paraprofessionals is extremely high—estimated nationwide at between 40% and 60% (Surpin, Haslanger, and Dawson 1994). The result is that many low-income women "cycle" between home care and welfare, entering low-barrier, low-quality employment as a home health aide and then—whenever the next family crisis necessitates—falling back to public assistance.

The Cooperative Enterprises

All three home care cooperatives discussed in this chapter act as subcontractors within their local health care markets. In many urban areas, a Medicare-certified home health agency—a visiting nurse association or a major hospital—will provide skilled nursing care and subcontract for paraprofessional services from an agency such as ours. In these cases, the homebound client will receive nursing services from the "professional" agency and aide services from the "paraprofessional" agency.

Remaining a subcontractor has obvious business limitations, yet creating a company that employs home health aides almost exclusively has allowed each enterprise to establish an extremely supportive "corporate culture" defined by and built around the frontline worker. The one-person, one-vote cooperative legal structure—with the vast majority of worker-owners being low-income women of color—in turn reinforces that culture.

Finally, the paraprofessional wages and benefits provided by the three cooperatives, although not as high as we would like, range from 10% to 20% higher than the norm within each local subcontractor home care market: Average wages range from $7.50 to $8.00 per hour, with hours per worker averaging between 31 and 34 per week. Individual health care insurance is offered to all workers who pass their probationary period.

Chronology

Cooperative Home Care Associates

In the early years, the survival of Cooperative Home Care Associates was by no means assured. CHCA was initially conceived by Rick Surpin and Peggy Powell from within the relatively safe Manhattan walls of the Community Service Society (CSS), one of New York City's largest nonprofit social service organizations. In the beginning, neither Surpin nor Powell expected to manage CHCA—they were economic developers who intended to use the nonprofit CSS as a staging area from which to launch a variety of worker-owned enterprises. However, after more than a year of chaos under an "industry-experienced" manager—during which the fledgling enterprise nearly went bankrupt—the two nonprofit developers soon found themselves managing day-to-day operations at the helm of their faltering for-profit venture.

The detailed story of CHCA's near-death experience has been eloquently recounted elsewhere (Dawson and Kreiner 1993). What is critical to note here is the entrepreneurial role played by the nonprofit CSS in starting and then nurturing CHCA through its early, troubled years—contributing not only Surpin's and Powell's time, but also investing grant support that substituted for equity. In this way, CSS's developmental capacity foreshadowed the creation of the nonprofit Paraprofessional Healthcare Institute and the Institute's entrepreneurial role in replicating CHCA in Philadelphia and Boston.

The Training Institute

In 1991, the nonprofit Institute was formed from the rib of CHCA, initially to run CHCA's entry-level training program, but soon to undertake the CSS-like entrepreneurial role of spawning new start-ups within other inner-city markets—providing professional expertise in the initial feasibility assessment, writing the business plan, hiring and orienting top management, and then sheltering the new enterprise with substantial equity and training funds.

The Replication Sites

In 1992, the Institute identified Philadelphia as the first replication site and hired Scott Gordon to be the CEO of Home Care Associates of Philadelphia.

With lead national support from the Charles Stewart Mott Foundation and the Ford Foundation, federal support from the Administration on Aging of the Department of Health and Human Services, and local support from the Pew Charitable Trusts, the Institute provided HCA substantial technical assistance for business planning and initial market development, invested $300,000 in equity, and arranged for $250,000 in long-term debt through a variety of program-related investments.

Viewed from a distance, the start-up of HCA appears to have been remarkably smooth. Under Gordon's leadership, the company broke even, on schedule, within 18 months of opening its doors. Within 3 years, HCA became the largest Medicare home care subcontractor in inner-city Philadelphia—based on its reputation of providing the highest-quality paraprofessional services in the city.

Furthermore, HCA has continued to build its workforce to more than 70 paraprofessionals—more than 85% of whom had formerly been on public assistance. HCA has also succeeded in upgrading the employment status of an additional 9 women who began with the company as home health aides and now work in administrative positions at HCA headquarters. Finally, more than 50 women have chosen to become worker-owners and have enjoyed dividends for the past 3 years averaging from $200 to $650 per worker.

Up close, however, the story of HCA is more complex, for the local home care market in Philadelphia proved far less fertile than that in New York: The average home care visit in Philadelphia is only 75 minutes, compared with New York's average of nearly 3 hours. Aides in Philadelphia must travel among several cases each day, with the result that hours per worker hover just above 30 hours per week—far above the local market's typical home care hours of 22 to 25 per week, but still significantly less than CHCA's average in New York of 34 hours per week.

Given this grudging environment, HCA has struggled since start-up to diversify its business services to provide more full-time work. In recent years this has led to a "temporary-to-permanent placement" strategy in which HCA trains and employs entry-level workers not only for its own home care business, but also for 3-month "temporary" placements within institutional settings, such as mental health facilities. Temporary employees who perform well are then hired by these new agencies, ensuring them full-time jobs.

On the one hand, this strategy promises a significant number of new, decent jobs—hopefully more than 50 per year by 1998. On the other hand, it challenges the cooperative culture of HCA, because successful employees will not stay as worker-owners of HCA but instead will become full-time employees elsewhere.

In Boston, in 1993, the Institute hired Seth Evans as CEO of Cooperative Home Care of Boston. Again, the Institute provided personnel, equity, finance, and training support identical to that provided to HCA in Philadelphia—this time assisted by the Mott and Ford Foundations, by the federal Job Opportunities for Low-income Individuals program of the Department of Health and Human Services, and by local support from, among others, the Boston, Hyams, and Riley Foundations.

Again, from a distance the start-up of CHCB appears nearly flawless: Under Evans's leadership the company broke even, on schedule, within 18 months and has remained profitable ever since. The Boston home care market's case configuration lies midway between those of the Philadelphia and New York markets—the typical visit averages slightly more than 2 hours—and with tremendous effort and ingenuity CHCB has been able to provide its workforce between 33 and 34 hours per week.

Furthermore, CHCB's contractors now consistently name the cooperative's services as among the highest quality in the city, and specifically ask for CHCB aides by name. Therefore, the workforce has grown now to more than 60 workers, and in early 1997 CHCB home health aides began purchasing their membership shares.

Yet viewed up close, the story is again more complex: In Massachusetts, welfare has been reformed with a vengeance, in particular disrupting the recruitment networks for new trainees that CHCB so carefully constructed during its first 2 years. Furthermore, CHCB has began to saturate its inner-city market: Additional growth of any significance will be obtained only if CHCB expands beyond Boston to cover surrounding cities. Although the resulting growth should provide the cooperative greater economic stability, clearly such regional expansion will, among other things, challenge the sense of community among the inner-city workforce that the cooperative forged during its start-up years.

Finally, in New York, Cooperative Home Care Associates is itself being forced to change: Hospital consolidation and changes in Medicaid regulations have combined in the past 2 years to cut the New York City home care market by 10%—but fortunately, close cooperation with the Visiting Nurse Service, CHCA's largest contractor, has recently placed the cooperative on a renewed track of job growth.

In addition, CHCA has embarked on an ambitious plan to create a specialized chronic care management organization for severely disabled children and adults. From an employment perspective, this initiative is designed to gain greater control over both reimbursement and service-provision mechanisms—allowing CHCA to strengthen the role of the paraprofessional and in turn create a better-trained, more highly valued, and better-paid career track for CHCA's home health aides.

Lessons of Replication

As the above chronology suggests, the replication of CHCA has little to do with transporting a static model into static sites. Instead, the replication of CHCA quickly became the introduction of a multifaceted, dynamic model into differing and very dynamic sites. Furthermore, the process is far from over: CHCA, HCA, and CHCB all continue to grow and, more important, all are substantially adjusting their initial strategies in response to the turbulence now churning throughout the nation's health care industry.

Organizational Characteristics

What remains constant within this complexity, however, are four essential characteristics shared by all of the three enterprises. First, they all possess an overriding *mission* to provide both high-quality paraprofessional jobs for inner-city women and high-quality care for clients who are elderly or disabled. Second, they all employ a set of organizational *core competencies* in the selection, training, supervision, and support of paraprofessional health care workers. Third, all cultivate an organizational *culture* that engenders openness, mutual respect, fairness, and a sense of community. Finally, all three possess an *organizational structure* that encourages workers' career development, participation in decision making, and ownership in their own cooperative.

Underlying Premises

In reflecting on this experience, we have confirmed some of our original premises. First, our ability to create better jobs for low-income people has been in direct proportion to how deeply we have been able to engage within local markets. As industry participants, we have access to information, opportunities, and, most important, relationships that are unavailable to even the most sophisticated researchers and analysts.

We have also confirmed that opportunities are created, not discovered. Our insider position has allowed us to identify and solve the problems of other industry actors in ways that meet simultaneously their business interests and our social goals. As a result, our intervention has constructed a pragmatic and compelling "business logic" for creating quality jobs for low-income people; we have forged an industry argument, not a charitable plea, for addressing a societal need.

Our replication program has also generated many new lessons that we had not originally foreseen. Three lessons have been key. The first involves CEO training and support. We miscalculated the relevance of CHCA's current experience to the reality of the start-ups: CHCA is a 13-year-old, 390-worker agency in a large and very established market, and the new managers were faced with the very different task of starting new agencies with very small staffs. On the one hand, this miscalculation generated a significant degree of frustration on the part of the new managers, who perceived that "New York" did not fully understand local realities. At the same time, the replication staff members were forced to provide far more on-site support—even to this day—than we had anticipated or were initially staffed to provide.

However, at least two initiatives undertaken to train and orient senior management were highly valued by the local managers. The first was a several-months-long "immersion" in CHCA's day-to-day operations—we invited prospective managers to work alongside various staff people in New York to see firsthand how various operations were handled, from training to case scheduling. The second initiative involved the selection of a replication staff person to work closely with each manager as a "mentor"—someone with whom the new manager could talk through problems on a regular basis and who could act as an advocate for the replication sites among the other New York staff.

The third lesson learned concerns the dynamics of replication. Replicating a successful program is far more complex, both substantively and psychologically, than traditional technical assistance. Emotional dynamics—of the founders wishing to protect the integrity of the original model, of new leaders wishing to create something new of their own—create a tension that appears to be endemic to the process. So far, the replication program has managed this tension—though not without considerable friction at times—through constant attention to communication and continuing attempts to restructure staffing and other resources to meet the changing demands of all three sites.

One structural change that has enhanced communication and created a greater sense of shared ownership within the Cooperative Network has been the placing of representatives from all three sites on the Institute's Board of Directors, which oversees the replication program. This formal representation has created a federationlike structure to our governance, which helps keep the Institute more accountable to the local sites.

Another structural issue is the composition of the management team. We initially envisioned a simple staffing model, with each replication site having a fully trained CEO who would in turn train other senior staff. We have since learned that we must attend directly to the training of middle managers as well—particularly the lead operations director and senior trainer.

In addition, we have found identifying women of color for the senior management position to be exceptionally difficult—of the three senior managers, two are white men and the third is a white woman—though all three sites have filled all other senior positions exclusively with women, most of whom are women of color.

Key Strategic Design Elements

Finally, each of the Cooperative Network's three enterprises has succeeded in conventional terms—as a profitable, for-profit business—but each has also succeeded in "social" terms—having benefited not only their low-income workers with higher-quality jobs, but also their clients with higher-quality care.

We believe our conventional *business* success is attributable to a set of "production elements"—including strong management, a demonstrated market demand, and adequate equity—that is now widely accepted as essential to any community-based enterprise initiative (Emerson and Twersky 1996). However, we strongly believe the Cooperative Network's *social* success is attributable to four strategic "design elements."

The Dual Model: For-Profit Business and Training Program

The Cooperative Network represents a dual model that integrates two distinct components: Each of the enterprises includes both a profitable business and an on-site, employer-based training program. The short-term classroom training program (ranging from 4 to 7 weeks, depending on the site) leads to immediate placement in a permanent, unsubsidized job within the cooperative. Then, once the new employee is on the job, the respectful style of management, in-service training, personal and vocational counseling, careful supervision, and (later) career upgrading programs together all weave for her a supportive work community and learning environment.

For the successful training participant, this dual model guarantees the availability of a decent job—secured through a series of small, structured steps throughout the often difficult transition from welfare to work. The process is nearly seamless as the participant moves from the training program into the enterprise, for the training program and the enterprise share the same mission, style of management, performance expectations, and even physical space.

For the cooperative, this dual model guarantees maximum control over recruitment, selection, and training, thereby ensuring the best employee performance possible—which the cooperative in turn uses to secure a high-end market niche.

A Sectoral Employment Strategy

The Cooperative Network model is a tested example of a *sectoral employment strategy,* that is, one that "targets a particular occupation within an industry, and then intervenes by becoming a valued actor within that industry—for the primary purpose of assisting low-income people to obtain decent employment—eventually creating systemic change within that occupation's regional labor market" (Clark and Dawson 1995).

Most industries currently accessible to inner-city, low-income women offer poor-quality jobs, with low pay, few benefits, and no chance of upward mobility. These entry-level jobs no longer act—as some once did for low-income people—as the "first rung on a ladder" leading to increased skills, responsibility, and compensation. This reality presents two strategic paths. The first is to identify industries that offer high-quality jobs currently inaccessible to low-income women, and then assist low-income women in securing those positions by eliminating barriers to employment. The second is to identify industries that already employ large numbers of low-income women, and then work to mold those poor-quality jobs into "decent employment." The Cooperative Network chose the second path: reshaping an industry that currently keeps large numbers of low-income women working, but poor.

Sectoral influence on an occupation can be achieved in two ways: by changing the public regulatory framework (e.g., through a "living wage" law that creates a wage floor for any occupation under public contract) or by changing private industry practice (e.g., through a labor innovation on the part of one competitor that is so compelling it forces other businesses within that market to respond in kind).

By intervening inside the home care industry as an employer, the Cooperative Network model uses both tactics: Within the regulatory framework, CHCA co-led a coalition of unions, consumer advocates, and service providers in securing labor reimbursement rate increases from the New York State Legislature. And within industry practice, the superior quality of the cooperatives has convinced contractors in each of the three local markets to place higher expectations on the labor standards of their other subcontractors.

A Labor-Based Design

The design of the Cooperative Network is intentionally labor based rather than neighborhood based. From an enterprise perspective this is essential, because a labor-based strategy recognizes that most businesses do not respect neighborhood boundaries—for customers, suppliers, or workers. Except for retail stores, businesses are not typically neighborhood phenomena, but regional phenomena.

From the community perspective, a labor-based strategy recognizes that, like everyone else, inner-city residents live within several overlapping yet distinct "communities"—neighbors, friends, family, church members, coworkers. Some of these communities are geographically based, some are not. Therefore, given modern-day transportation and communication, the Cooperative Network model assumes that community should not be perceived solely in neighborhood terms.

In low-income communities, neighborhood strategies and labor strategies can be mutually reinforcing—which is why all three cooperatives work in partnership with local community development corporations. As those geographically based organizations work to strengthen the bonds of community among groups of neighbors, the cooperatives work to strengthen the bonds of community among groups of coworkers. Clearly, helping Latina women who live in one low-income neighborhood to develop mutually supportive work relationships with African American women who live in another is one effective way among many to "build community."

However, although all three cooperatives work in partnership with local neighborhood groups, the Network has intentionally avoided creating an ownership structure that blends worker and neighborhood control, believing that the political complexity of such a dual ownership structure would entangle what must remain an agile, market-oriented enterprise.

Perhaps most important, the experience of the Cooperative Network enterprises has now provided incontrovertible proof that this type of service business can be managed to "maximize the value of labor," and yet still succeed within a market economy.

Any successful business within a market economy must manage several factors simultaneously—return on capital, cash flow, labor, technology, and market position, to name a few. However, closer inspection reveals that businesses are often managed not from a logic of equalizing all factors, but from a logic of maximizing from among them one key factor, with the others managed to ensure the primacy of that key factor. The choice of which factor to maximize is to some extent dictated by the structure of the particular industry, but also—and this is critically important—it is determined in large part by the values of the decision makers in a particular business. For example, corporate investment firms often manage their holdings to maximize short-term return on investment, and in the process are accused of ignoring the harm their corporate decisions place on other "factors," such as consumers, workers, and communities. McDonald's manages for market share, pouring hundreds of millions of dollars into advertising. The local auto repair shop manages for cash flow, basing almost all business decisions on how cash can best be conserved. Apple Computer was maximized for technology—to the eventual detriment of its market share.

All of the businesses in these examples—with the apparent exception of Apple—are able to thrive in the marketplace while maximizing different variables. What the Cooperative Network has proven is that business can also be managed to maximize *labor* and nonetheless remain profitable within the marketplace. Should all of us who are working to create community/labor-based businesses ever demonstrate this same phenomenon in a wider range of companies within a greater variety of market settings, the implications would be profound for the building of a more humane, labor-based market economy.

One example of maximizing labor from within our cooperatives may prove helpful: As I noted earlier, turnover in paraprofessional positions in the home care industry is extremely high—ranging between 40% and 60%. The cost of replacing a worker—including recruitment, training, on-the-job orientation, and increased supervision—is approximately $3,500 per new employee. Therefore, a 100-worker agency spends nearly $175,000 every year ($3,500 times an average of 50 new workers annually) simply to maintain its workforce.

Clearly, a home care agency that invests in its frontline workforce will save substantial money if the result is lower turnover. Yet, realistically, investment in the frontline workforce *costs* money as well (in higher wages and benefits, for example). Therefore, both high-turnover and low-turnover agencies might in the end achieve the same profitability, but the low-turnover (high-investment) agency will have created a higher-quality service and a far more humane company. Furthermore, as has been true for all three of our cooperatives, the resulting high-quality reputation may in turn be rewarded in the marketplace through the generation of increased demand for services.

The Cooperative Network invests in its frontline workforce because the frontline worker is at the core of the Network's mission—a mission reinforced by the fact that the owners of the companies are the frontline workforce. Thus we are managing the companies primarily to maximize labor, yet when such management has been undertaken thoughtfully, it has proven in addition to be a successful, profitable business strategy.

Conclusion

Today, the Cooperative Health Care Network includes three home care cooperatives. For all their differences of size, market, and leadership personalities, they are remarkably similar—both in their day-to-day functions and in their core mission. However, 3 years from now, the Network will appear increasingly different: The sites will continue to evolve in directions that meet the varying demands of their local markets and the visions of their local leadership teams. In addition, the sheer number of entities associated with the Network will grow: CHCA in New York will initiate a separate chronic care management organization, HCA in Philadelphia will create its own distinct training and enterprise development corporation, and CHCB in Boston will likely open branch offices in surrounding towns. The Institute will also add entirely new enterprise sites as well—each of which will no doubt, in this chaotic health care industry, be uniquely configured to meet local market conditions.

At that time, we will judge our success by how well we have maintained throughout the Cooperative Network our core mission: maximizing the value of the frontline worker to provide high-quality care—in turn creating entrepreneurial, dynamic enterprises.

References

Bayer, E. J., R. I. Stone, and R. B. Friedland. 1993. *Developing a caring and effective long-term care workforce.* Menlo Park, CA: Project Hope Center for Health Affairs, Henry J. Kaiser Family Foundation.

Clark, P., and S. L. Dawson, with A. J. Kays, F. Molina, and R. Surpin. 1995. *Jobs and the urban poor: Privately initiated sectoral strategies.* Washington, DC: Aspen Institute.

Dawson, S. L., and S. Kreiner. 1993. *Cooperative Home Care Associates: History and lessons.* New York: Home Care Associates Training Institute.

Dow, M. M. 1991. *Managed care digest, long-term care edition.* Kansas City, MO: Dow.

———. 1993. *Managed care digest, long-term care edition.* Kansas City, MO: Dow.

Emerson, J., and F. Twersky. 1996. *New social entrepreneurs: The success, challenge and lessons of non-profit enterprise creation.* San Francisco: Roberts Foundation.

Himmelstein, D. U., J. P. Lewontin, and S. Woolhandler. 1996. Medical care employment in the United States, 1968 to 1993: The importance of health sector jobs for African Americans and women. *American Journal of Public Health* 86(4).

Lehmann, N. 1994. The myth of community development. *New York Times Magazine,* January 8.

Nye, N., and R. Schramm. 1994. *Building a learning organization: Final evaluation report.* Plainfield, VT: Goddard College Center for Business and Democracy.

Surpin, R., K. Haslanger, and S. L. Dawson. 1994. *Better jobs, better care: Building the home care workforce.* New York: United Hospital Fund.

Vidal, A. 1992. *Rebuilding communities: A national study of urban community development corporations.* New York: New School for Social Research.

8

Ready for Work

The Story of STRIVE

LORENZO D. HARRISON

Current welfare-to-work efforts have focused on improving the "soft skills" or work ethics of job seekers, and on linking them rapidly to jobs or work first, rather than providing long-term skills training (Brown 1997; Wilson 1996). An employment program that combines these two elements and adds employer connections, credible staff and trainers, and postplacement supports is STRIVE, incorporated as East Harlem Employment Service in New York City. From an organization that made 56 placements with a staff of only three in 1985, STRIVE has grown to an organization with a budget of $3,139,511 in 1997, and has placed and retained well over 17,000 underprivileged job seekers in New York and at other sites. From one office in New York, STRIVE has expanded and replicated to include 12 sites in New York City, 4 in Pittsburgh, 2 in Chicago, and single sites in Boston and Philadelphia.

The goal of STRIVE, has been to create long-term labor attachment for low-income people: to help participants acquire a solid work ethic, to place 80% of graduates in jobs, and to retain 80% of these in jobs over the subsequent 2 years. Although STRIVE was not initially explicitly concerned with changes in the income of participants, assuming that long-term retention in jobs would produce these benefits, the program's leaders recently realized that some participants were getting stuck at certain job levels. Consequently, STRIVE has turned its attention to career ladders and ongoing training. During its evolution, STRIVE has received recognition by the news media, government, and peer employment placement organizations (U.S. Department of Labor 1995; Hymowitz 1997a, 1997b; Rosin 1997). In May 1997, *60 Minutes* featured STRIVE as a promising employment training model.

This chapter tells the story of STRIVE's development from the perspective of the practitioners who made it happen and illustrates how the organization has continuously reflected on the basics of its employment approach throughout its history. This constant reflection has led the program to become more concerned with how to sustain people on jobs, how to enable them to obtain better jobs over time, and how to replicate the STRIVE model in different community contexts. I conclude the chapter by reviewing evaluation research on STRIVE and the lessons of this program for jobs and welfare-to-work programs.

The History of STRIVE

The Early Years (1984-1989)

In 1983, Samuel A. Hartwell, a corporate businessman and avid volunteer for social causes, had a notion about assisting individuals from poor communities in getting jobs. Up until that time, he had been a trustee of numerous nonprofit human service agencies and had become frustrated by how these efforts failed to help underprivileged young men and women overcome social and economic injustices. Hartwell thought that of all the social ills afflicting people who live in poor communities, unemployment and economic insufficiency stand out as the most profound. His first observation was that the problem of unemployment for the disadvantaged was one of "access."

In December 1984, Hartwell and several colleagues incorporated East Harlem Employment Service/STRIVE as a 501(c)3 nonprofit organization with a mission to address the chronic unemployment in the Harlem community and East Harlem in particular. The initial design of the STRIVE program built upon job readiness tenets and interpersonal skill development, bridging the void that Hartwell characterized as access. Lyle Gertz, an employment training specialist, became the first STRIVE staff person and invented the acronym STRIVE, which stands for Support and Training Results in Valuable Employees. For a time,

Hartwell and Tom Rodman, a partner at the financial brokerage house of Alex Brown, paid Gertz's salary out of pocket.

The focus on access shaped the initial STRIVE program as short-term, based on solid relationships with private sector employers. STRIVE viewed employers as customers, and attempted to reduce their turnover and human resource costs. If participants of STRIVE training lacked connections through their parents or aunts and uncles, then STRIVE was going to step in and provide those connections for them. In May 1985, on a budget of $120,000, STRIVE enrolled its first group of participants. At the start, the program offered no stipends for participants, advertised in the jobs section of the *New York Daily News,* and stayed with each graduate regardless of how many interviews he or she needed to get a job.

During the first class at STRIVE, the talents of Frank Horton emerged; he would later become STRIVE's director of training and technical assistance. Horton possessed employment experience, had experienced New York's foster care system, and had successfully completed the 3-week employment training workshop. He was hired as the second STRIVE staff member in June 1985.

Horton contributed to the design of STRIVE's training approach—an approach that taught participants how their already existing intangible survival skills were transferable to entry-level employment. Throughout 1985-1986, Gertz and Horton, under the leadership of the Board of Directors, continued to develop various world-of-work and life skills training techniques to assist young Harlemites in accessing jobs. STRIVE survived the growing pains that any start-up company goes through while achieving 56 job placements that first year. STRIVE graduates filled positions as assistant porters, laborers, mail room clerks, telemarketing agents, hospital patient escorts, and office receptionists.

The board conducted a job search for STRIVE's first executive director in 1986, and eventually hired Robert Carmona, who brought to STRIVE a diverse background that became a standard for STRIVE staff. Before coming to STRIVE, Carmona, a graduate of Columbia School of Social Work, had served as a youth counselor at Downstate Hospital in Brooklyn, as a family therapist at Daytop Village, as a funding officer at United Way, as a coordinator at City Volunteer Corporation, and as a director of marketing at Wildcat Service Corporation. In addition, Carmona had overcome the travails of growing up on the Lower East Side of Manhattan and Staten Island. At the same time Carmona was hired as executive director, the organization hired Lawrence Jackson as a job developer. Jackson was a native son of the James Weldon Johnson Housing Project, where STRIVE was located.

Attitudinal Training: The Cornerstone of the Curriculum

In the spring of 1987, STRIVE began to describe its program with a new phrase: *attitudinal training*. Although the principals of STRIVE understood the importance of incorporating job readiness into the program, the techniques of job readiness did not adequately address an issue that the STRIVE staff found prevalent among its clientele: poor attitudes. Most of STRIVE's participants were African Americans and Latinos between the ages of 17 and 40. The profile of STRIVE participants ran the gamut: former substance abusers, high school dropouts, single mothers receiving public assistance, ex-offenders, and youngsters phasing out of the foster care system. Common job readiness training did not meet all their needs. It covered such topics as how to fill out job applications, the correct format for a résumé or cover letter, the importance of appropriate business attire, interviewing skills, the value of punctuality, and other necessary but superficial elements involved in preparing oneself for work. Furthermore, its teaching methods were pedantic.

STRIVE developed a quite different approach. As an illustration, a 22-year-old man who was an aspiring member of a Bronx youth gang came to STRIVE and filled out an application at the direction of his probation officer. In observing the young man's physical demeanor and manner of speech, it was not difficult for the staff to see that he had a "tough guy" persona, negotiating life through intimidation. During the course of the orientation, Carmona and Horton singled out his attitude from among the rest of the group. They asked the young man what he was trying to prove. Did he know any gangsters who lived past the age of 35? Why was he there?

This type of quasi-confrontational group dynamic became STRIVE's hallmark. Attitudinal training demands self-examination, critical thinking, relationship building, and affirmation. It attempts to cut through the emotionally laden "baggage" that many inner-city residents use to navigate life. STRIVE's training approach is particularly effective because the organization ensures that the trainers themselves come from lived experience similar to that of participants—many of them are STRIVE graduates.

Similarly, STRIVE's training confronts racism in a pragmatic way. For example, a trainer may jar the thinking of the group by telling a true story: In 1987, an African American female STRIVE graduate who worked as part of an office support pool at a communications company in New York City called to say that she was becoming extremely offended by her supervisor's racist actions. As part of a team of six office assistants, she was the only black in the pool and one of few blacks in the entire company. Her supervisor, who was white, gave her disproportionately more work than her white colleagues. STRIVE investigated and found out that the reason this woman's supervisor gave her most of the work was because she was the best worker on the team.

Another approach trainers use is to pose a series of rhetorical questions: If you apply for a job and your competition is a white person and all things are equal, do you think you will get the job? Answer: Maybe, it depends. Another question is posed: Is racism still alive in America? Answer: Yes, it is. Final question: At the absolute end of the day are you responsible for your disposition in life? Answer: Yes, resoundingly! STRIVE forces participants to examine their own perceptions, attitudes, and mind-sets toward work, responsibility, and citizenship.

STRIVE sees attitudinal development as the lever or "catalytic converter" that when triggered can turn a young man or young woman around. STRIVE's attitudinal training demands that participants take ownership of their attitudes. "Taking ownership" challenges participants to look within themselves and own up to their mistakes in life. In this way, STRIVE's training forces participants to understand that changing their disposition in life is within their control; consequently, participants begin developing a sense of empowerment. This approach has been described as a mental boot camp for the "hard to employ": It's a no-nonsense, hard-driving approach, comprising 120 instructional hours. The program takes place in a simulated work environment where excuses are unacceptable and participants are provided the tools they need to begin viewing themselves as marketable adults.

In today's service-oriented job market, entry-level candidates need to be self-motivated and must possess strong interpersonal skills. STRIVE's training is based on the premise that many personnel directors and employment coordinators will take a chance on a prospect who exhibits not only the necessary aptitude, but the "right" attitude for the entry-level work environment. Part of what employers want is for employees to value a "spirit of cooperation" and to be able to accept constructive criticism; STRIVE seeks to develop this spirit of honesty and teamwork among participants.

Replication of the STRIVE Model

As the STRIVE methodology solidified, Carmona hired another important staff person in the evolution of STRIVE. Toward the latter part of 1987, Carmona began discussions with a young man he had met some 11 years prior when they were "squaring up," as it's said, in upstate New York at Daytop Village, a confrontational drug treatment program. Carmona knew that Lorenzo Harrison held a strong belief and conviction, like himself, that people who are labeled or stigmatized as being from the so-called underclass in fact have the ability to turn their lives around. At the time Harrison was at a crossroads in his life after becoming disillusioned with the complexities of divestiture during a 5-year tenure at New York Telephone Company. Carmona convinced his friend to take the part-time assignment as a job developer on a STRIVE-related program.

STRIVE's maturation brought with it the opportunity and challenge of replication. In late 1987, STRIVE began discussions with groups in Pittsburgh about establishing a STRIVE company. Hartwell hailed from an old Pittsburgh family and belonged to a venture capital group based in Pittsburgh. He and his business associates polled constituents from the human service community, elected leaders, government officials, corporate heads, and foundation officers to gauge the feasibility of a STRIVE program in the city of Pittsburgh.

Shortly after that, in 1988, they secured financial commitments for the first year's budget for Three River Employment Service Inc./STRIVE. Simultaneously in New York, Jim Canales, the executive director of what was then called the Bronx River Neighborhood Center, Inc., reached out to STRIVE to learn more about its employment approaches and encouraged Carmona to consider collaboration. The Clark Foundation, which had provided seed money and ongoing support to STRIVE since 1985, was also a funder of the work of Bronx River Neighborhood Center. Joseph Cruickshank, a senior program officer at the Clark Foundation, along with principals from STRIVE's board and Canales, enabled Bronx River to incorporate a STRIVE component at its settlement house.

Long-Term Follow-Up

As new participants were being trained and placed every month, Carmona and his staff realized that approximately 15-17% of the individuals who were served in the latter part of 1985 and the beginning of 1986 had returned to the agency seeking additional support services. Some of these individuals had been separated from the initial jobs in which they were placed and returned to job searching. As a result of the recognition of the postplacement needs of its graduates, STRIVE instituted a follow-up system called Graduate Services. Though STRIVE had long identified itself as a lifetime service to which participants were always welcome to return, it now determined that dedicated follow-up was necessary. In developing the follow-up structure, STRIVE realized that the 90-day Job Training Partnership Act criteria would be insufficient.

The STRIVE follow-up system involves daily, weekly, and quarterly contacts with all graduates, and includes evening and weekend activities, an alumni association, re-placement and upgrade services, counseling, and referrals for other support services, such as for continuing education and problems with housing, domestic abuse, and child care.

In addition to meeting the many ongoing needs of STRIVE graduates, another important aspect of STRIVE's follow-up system is that it measures and tracks the retention of STRIVE graduates in the labor market. Simply stated, STRIVE's programmatic goals are to place 80% of its graduates and to retain 80% of those placed in employment. STRIVE has been deliberately molded to be a results-oriented organization. Its dedication to the postplacement development of its clientele, and the prerequisite assessment, analysis, and evaluation, make the STRIVE name synonymous with attitudinal training and long-term follow-up.

The Clark Challenge (1990-1997)

On December 6, 1989, the Clark Foundation—a consistent longtime supporter of STRIVE—awarded STRIVE a multimillion-dollar challenge grant to establish a network of STRIVE programs throughout New York City. STRIVE was ecstatic at this news, but there were also questions: What would this mean for a mom-and-pop community organization? Would it cause STRIVE to revisit its mission? With its newfound status as a multimillion-dollar agency, would STRIVE have to abandon East Harlem? How could this expansion and replication be managed properly? With these questions and others confronting the agency, board and staff recommitted to STRIVE's basic mission: to prepare, train, place, and support inner-city youth and young adults in long-term employment.

STRIVE's board, staff, and funders tackled the challenge of designing the organizational structure of the network, which would become the STRIVE Employment Group (SEG). Issues centered on whether the network would be incorporated separately from East Harlem Employment Service/STRIVE or implemented as a component under its existing corporate charter. One major condition of the Clark challenge grant was that for every two dollars raised by STRIVE, the Clark Foundation would contribute one dollar. The organizing team explored the feasibility of involving existing nonprofit agencies that were interested in STRIVE's having the ability to raise these funds. The trustees of the board and staff of the Clark Foundation, Carmona, and Michael Frey, a STRIVE board member, worked to negotiate arrangements that did not require prospective partners to raise new moneys or require private dollars to match the challenge grant. The important initial conditions of STRIVE's employment network were that STRIVE would function as the managing and fiduciary agent for the Clark funds, that the entity would not be separately incorporated, and that prospective partners could utilize in-kind contributions to match the Clark challenge grant.

Of course, the plans were not perfect. The Clark Foundation originally stipulated that the $4 million challenge grant would be disbursed at $1 million per annum from January 1990 through December 1993, with a goal of placing 10,000 disadvantaged New Yorkers in unsubsidized entry-level employment. The difficulty in meeting that goal related to the challenge of building relationships with colleagues who would form the network. Again, Clark was flexible in relaxing that condition of the grant to allow STRIVE as much time as necessary to develop quality relationships. STRIVE solidified five collaborative agreements in 1990—new sites in the North Bronx, the Lower East Side, Midtown and Jamaica, Queens. The collaborative agreements included a diverse group of settlement houses, local development corporations, and foster care agencies that focused their services on ex-offenders, former substance abusers, and welfare recipients. The partner agencies would implement, establish, and develop STRIVE component programs at their existing locations. STRIVE understood that partner agencies would have to be genuinely committed to embracing STRIVE's approach to employment, given their neighborhood and organizational context. This was not to be cookie-cutter replication.

Refining the Basics

As STRIVE's caseload of graduates continued to grow each month, it became difficult to sustain quality follow-up for each graduate. STRIVE struggled through this learning period while its leaders engaged in discussions with colleagues and other practitioners about providing follow-up services and a postplacement safety net to enhance the chances of graduates' maintaining employment. In conversations with colleagues in government, STRIVE's leaders learned that managers within the welfare department would not totally close the books on recipients until they were separated from the welfare rolls for a minimum of 2 years. They seemed to have determined that the rate of former recipients' returning to the rolls was very minimal if they remained in the labor market for approximately 2 years.

STRIVE staff did some of their own in-house research and analysis, and determined that graduates who were in the labor market for roughly 2 years experienced few additional periods of unemployment. As a result, STRIVE redefined its programmatic structure for follow-up; it would still provide lifetime services involving job placement, skill upgrading, counseling, and training, but after 2 years of the date of placement of a graduate on a job, the onus would be on the graduate to return to the agency if he or she needed additional services. In this way, STRIVE remained firmly committed to the long-term development of all its graduates while simultaneously remaining cost-effective.

Member agencies appreciated the minimal paperwork and administrative tasks associated with participating in the STRIVE network. One parameter that the STRIVE management team sought to maintain was that historically the STRIVE program had per capita costs of approximately $1,200 and $1,500 for placement and retention of participants, respectively. STRIVE managers and staff implemented a strategy for providing technical assistance to each of the partner agencies.

In addition to the ongoing development of the STRIVE Employment Group, Carmona and Harrison reaffirmed senior management's commitment to STRIVE's programmatic basics. They asked new questions: Why was it so difficult to get as many men in the program as women? If we reviewed the retention information in a different way, could it tell us more about our graduates? If we measured how long participants stayed on particular kinds of jobs, could that tell us something new about the labor market? What percentage of working STRIVE graduates are truly self-sufficient? Are there discernible occupational trends for STRIVE graduates, and, if so, where do those trends lead? Are there other correlations that could inform program development? Given STRIVE's no-nonsense approach to employment preparation, how do we improve the extent to which we toe the thin line between what's perceived as tough love and callousness? With these kinds of questions serving as challenges, Carmona, Harrison, and other staff sought to improve STRIVE's practices and techniques for assisting inner-city youth and young adults.

Replication in Chicago built upon the experience of start-up in Pittsburgh, and paid attention to local context and constraints. Over time, the Chicago site modified the STRIVE training model to be less confrontational because the program drew an older clientele with more work experience. The site had experienced much higher dropout rates using the original model. In the fall of 1994, Boston Employment Service/STRIVE was incorporated and established in the Dorchester community of Boston, Massachusetts. STRIVE's replications have always started out with small staffs and modest goals. The aims are for a start-up company to stabilize its practices, firm up its funding, and find its identity. Once these tasks are accomplished, it is easier to build capacity to full scale. STRIVE never attempts to serve more than 300 individuals per year,

25 a month, at any given site. Experiments in New York have proven that anything above that plateau changes the chemistry in respect to the staff:participant ratio.

Moving Beyond Placement to Career Development

At the board retreat in September 1993, STRIVE directors and managers became concerned about information they were receiving relating to graduates who were "maxing out" of the 2-year follow-up and still in need of support. Was there a fundamental flaw in STRIVE's premise that once individuals were in the labor force for a period of 2 years, they had acquired the wherewithal and tools to sustain themselves, to attain self-sufficiency and independence?

What the board and senior management realized after months of research and analysis was that there was not a fundamental flaw in the program's conceptual base, but there was a gap. Although many graduates proceeded to build on their STRIVE experience by returning to the agency for assistance with job upgrades, lateral moves, and career advancement counseling, there was a segment of STRIVE's caseload that, for various reasons, needed a structured approach to assist them with upward movement in the labor market.

One example given was that of a 24-year-old man who maintained a shipping and receiving job for 2 years after graduating from the workshop but was making only 50 cents more than the $6.00 per hour wage he received when he started and saw no opportunity within or outside the company for advancement. He was still living at his aunt's apartment in the Bronx and could barely afford to take his girl out to the movies on the weekend. Could STRIVE count the development of this young man as a success? He worked for the 2 years and seemingly had acquired a solid work ethic. He participated when he could in some of STRIVE's postplacement activities. Based on this and the experiences of other participants, the board and staff began to design a career advancement strategy.

As with the STRIVE network, this effort took on several incarnations. First, Harrison and Michael Blum, a long-standing STRIVE board member from Coopers and Lybrand, devised the Graduate Career Development Program—a strategy to afford STRIVE graduates the opportunity to create professional career road maps. The program never got off the ground due to a lack of funding and staff resources; however, Chicago STRIVE experienced some success in launching a career development vehicle called Career Path.

At the same time, Carmona became reconnected to an old acquaintance, Dr. Ronald Mincy, now a senior program officer at the Ford Foundation. After learning about STRIVE's ambition to construct an initiative to move its graduates into better jobs, Mincy envisioned a collaborative arrangement that would become known as the Access, Support and Advancement Partnership (ASAP). Beginning in late 1994 and early 1995, Ford enabled STRIVE and its collaborators, Stanley M. Isaacs Neighborhood Center in New York and Jobs for Youth in Boston, to plan the ASAP program for 18 months. The conceptual framework for the effort built upon the employment access, support, and advancement strengths of the three partners. STRIVE had made its name by being able to move "hard-to-employ" individuals into jobs relatively quickly. Stanley Isaacs Center had a reputation in the New York City human service community for its strong commitment to youth development. Jobs for Youth's forte was in the environmental technology and biomedical technology, in advanced placement, technical skills training, and identification of training niches.

On October 1, 1996, STRIVE launched the Access, Support and Advancement Partnership—a $4.2 million multiyear career development program for STRIVE graduates. ASAP assists employed STRIVE graduates in advancing in the labor market and acquiring positions that pay livable wages in growth industries. Among the most intriguing aspects of ASAP are its eligibility criteria: Participants have to be graduates of STRIVE's baseline program in New York or Boston (or, in the case of Jobs for Youth, its job readiness equivalent) who have worked for at least 8 to 12 months. ASAP provides training during evening and flex hours to enable participants to maintain their current employment. The ASAP team has ascertained economic self-sufficiency in New York to be an annual salary of $23,000, and in Boston, an annual salary of $21,000.

Evaluation

STRIVE has never been the subject of a random assignment training evaluation, nor has it had the benefit of evaluations that utilize comparison groups or quasi-experimental designs. Yet there have been numerous evaluations that demonstrate the merit and cost-effectiveness of short-term job search and readiness programs (U.S. Department of Labor 1995). New York University's Wagner School of Public Service conducted the most thorough study of STRIVE for the original East Harlem office in 1993-1994. The aim of this study was to provide recommendations for improving management practices (New York University 1993). In conducting the study, the researchers looked at performance data and at STRIVE's recruitment, outreach, postplacement, and training systems, and interviewed a small sample of participants and employers.

The study found that STRIVE had achieved a 75% placement rate and an 82% retention rate between 1985 and 1993, compared with STRIVE's goal of 80% for both placement and retention. Of the participants studied, 63% were women and their median age was between 20 and 25 years. Receptionist/clerical and cashier positions accounted for most of the entry-level placement positions. The study found also that STRIVE experiences two periods of participant attrition: during and after the orientation session, when 40% drop out, and during training, when 5% drop out. Sampling of dropouts by STRIVE suggests that as many as one-third may eventually reenter the program.

STRIVE, has also been the subject of several national "best practice" reviews of the most effective and promising employment and training programs. A U.S. General Accounting Office study in May 1996 included STRIVE along with the Center for Employment Training and Focus Hope of Detroit. STRIVE was also highlighted in the recent U.S. Department of Labor and Ford Foundation PEPNet (Promising and Effective Practices Network), a certification study of effective youth employment programs. The aims of PEPNet were to establish high quality standards and to identify organizations throughout the country that met them. Again, STRIVE stood out as one of 18 national award winners (National Youth Employment Coalition 1996).

In October 1995 it happened: The STRIVE Employment Group made its 10,000th placement of a young person in New York City. This remains one of STRIVE's proudest achievements. More important, of the 10,455 placements by that year's end, SEG staff determined that almost 77% of the individuals placed were still in the labor force.

Summary of STRIVE's Lessons Learned

STRIVE views itself as an unorthodox employment preparation program that does not necessarily comply with the assumptions of traditional employment training and placement programs. The most unique aspect of STRIVE's service delivery is its focus on attitudinal training and development. STRIVE has learned over the past 13 years that employers hire its graduates because they are "attitudinally correct," conscientious, likable, intelligent, and ready to learn and work; they have good interpersonal skills and appear to have made some concessions to mainstream social norms.

Although some practitioners are adding soft skill components to their overall program designs, they face two challenges. One is that many of them appear to go only part of the full distance in incorporating comprehensive soft skill training curricula that reach below the surface and address the perceptions, attitudes, and emotional makeup of their participants. The second is that staff tend to be more effective when they can relate to the experiences of participants. STRIVE has determined that its capacity to "deliver" on an attitudinal level is tied directly to the fact that most of its staff members are individuals who themselves have overcome the same life-threatening hardships as their clients.

STRIVE believes that participants are the ultimate determinant of whether or not employment programs achieve their goals. Although STRIVE utilizes standard assessment instruments to assess participants, its most important tool is the dynamic process by which it assesses the attitudinal development of workshop participants. The most important objective of the orientation is to ascertain participants' sincerity and drive to "turn their lives around." Candidates for the STRIVE training workshop must indicate that they are genuinely interested in improving themselves in regard to their employability. STRIVE's training and clinical staff have honed their abilities to assess subtle behaviors that reflect these underlying attitudes. Professionals are unable to fabricate within an individual any inner will, desire, or ambition. However, skilled and dedicated professionals are able to build upon the smallest evidence of such feelings.

Conventionally, employment training and placement programs are measured by the extent to which they are able to keep participants. Although there is certainly a legitimate value to this measure, it is too often compounded by externalities outside the control of service providers. Ultimately, STRIVE believes that most of the young men and women coming through its doors are prepared to invest in themselves, and that, further, the prospect of that self-investment is significantly enhanced when the service is of quality and the service providers are sincere.

The second lesson that STRIVE has learned is the importance of maintaining contact with its graduates long after the time they are placed on a job. Early in its development, STRIVE was challenged to explain how its program was more than just a 3-week formula. STRIVE answered that it believed that the problem of employment for young men and women from poor communities was one of access. Further, the best training for work is work itself. STRIVE explained that it offers a lifetime service for graduates encompassing follow-up contact, counseling, evening and weekend activities, re-placement and upgrade services, and referral services. In fact, a predictable percentile of STRIVE participants will lose their first jobs. STRIVE's approach does not view this as failure, but rather as part of clients' ongoing learning experience and development. The goal is to assist participants in acquiring a solid work ethic. STRIVE's follow-up system provides a safety net that intervenes, retrains, re-places, and reengages graduates at critical junctures throughout their development.

The Access, Support and Advancement Partnership, a comprehensive career advancement program, could prove to be STRIVE's most important lesson learned. The goal is not simply employment, but rather self-sufficiency. The two main conditions of eligibility for the ASAP program are that a person must be a STRIVE graduate and must have worked for at least 8 to 12 months post-STRIVE. Building on the unique assessment and eligibility criteria of STRIVE's baseline model, ASAP demands that the STRIVE graduates demonstrate their confidence, self-respect, and continued motivation in the labor force by working for roughly a year. Given this prerequisite, STRIVE believes that its services will be invested in individuals who are sincere.

Finally, STRIVE believes, as a result of its replication activities, that successful employment training initiatives have to be based on strong community involvement. This is demonstrated by the development of STRIVE's original site in East Harlem and its community ties. Moreover, all of STRIVE's replications developed genuine community support before they were launched. East Harlem Employment Service/STRIVE expects to launch a significant initiative with formal links to East Harlem Partnership for Change, a grassroots community organizing campaign affiliated with the Industrial Areas Foundation, that could have national implications in terms of promoting accountability in government-funded employment programs.

STRIVE developed in unforeseen ways. It was not planned, or predetermined, that STRIVE would one day be viewed by some as a leader in the employment training and placement arena. However, what has become very clear is that an originally small and focused community organization can have far-reaching influence.

References

Brown, A. 1997. *Work first: How to implement an employment focused approach to welfare reform.* New York: Manpower Demonstration Research Corporation.

Hymowitz, K. S. 1997a. At last, a job program that works. *City Journal,* Winter.

Hymowitz, K. S. 1997b. Job training that works. *Wall Street Journal,* February 13.

National Youth Employment Coalition. 1996. *PEPNET '96: Promising and Effective Practices Network.* Washington, DC: National Youth Employment Coalition.

New York University, Robert F. Wagner Graduate School of Public Service—Management Clinic. 1993. STRIVE's results: Evaluating a small non-profit organization in East Harlem. Unpublished manuscript, New York.

Rosin, H. 1997. About face. *New Republic,* August 4, 16-19.

U.S. Department of Labor. 1995. *What's working (and what's not): A summary of research on the economic impacts of employment and training programs.* Washington, DC: U.S. Department of Labor.

U.S. General Accounting Office. 1996. *Employment training: Successful projects share common strategy.* Washington, DC: U.S. General Accounting Office.

Wilson, W. J. 1996. *When work disappears: The world of the new urban poor.* New York: Knopf.

9

New Careers Revisited

Paraprofessional Job Creation for Low-Income Communities

JANICE M. NITTOLI
ROBERT P. GILOTH

A perennially attractive strategy for improving the economic self-sufficiency of poor communities is the utilization of the skills and capacities of indigenous residents of these communities to deliver social, educational, and health services. Those who deliver these services are referred to by many names, including *paraprofessionals, outreach workers,* and *lay advisers.* Not surprisingly, this approach is relevant today as well, as cities and states gear up to implement changes in welfare, with a premium placed on work. Yet the track record of paraprofessional programs in creating careers and family-supporting jobs for low-income residents raises many questions.

Interest in enlisting low-income residents into paraprofessional occupations has at least three rationales. First, millions of dollars are spent annually in low-income communities to solve "human service" problems—child welfare, juvenile justice, health, housing, and income supports, to name a few. However, few of these dollars employ residents of these same communities. A recent study of the Grand Boulevard neighborhood in Chicago, which contains the Robert Taylor Public Housing Project and encompasses some of the poorest census tracts in the country, estimated that $167 million were spent in 1994 in a community of 36,000 residents. The magnitude of this public investment raises the question of whether these dollars could be spent to create jobs and economic opportunity while contributing to the solution of human service problems—what some call *double social utility*. Analysts have calculated that this Grand Boulevard investment translated into 634 jobs, more than one-third of them aide jobs and 41 jobs at higher skill levels (Bush, Ortiz, and Maxwell 1995).

The second rationale for the support of paraprofessional programs is more effective human services. Those millions of dollars spent in low-income communities seldom solve the problems for which they are intended because the programs they fund are often fragmented, bureaucratic, disempowering, or culturally incompetent. Indigenous workers, as opposed to outsiders, it is believed, have the ability to reach community residents, to communicate and empathize within specific contexts, and to deliver services in a manner that is likely to be better understood (Kinney et al. 1994; Larner and Halpern 1987; Larner, Halpern, and Harkavy 1992b; Bruner 1996).

The third rationale is that paraprofessional jobs place a premium on local rather than certified professional knowledge. Hence these jobs are accessible to people with few formal skills and do not have insurmountable barriers to entry. They take advantage of informal skills and life experiences. Yet these jobs also have the potential to serve as entry points into careers that, in the long run, offer higher levels of compensation and responsibility (Pearl and Riessman 1965).

Today, two additional reasons exist for renewed interest in paraprofessional jobs. As mentioned, changes in welfare put a premium on work and are the responsibility of states—also the funders of many human services. In many cities and rural areas, lack of jobs in the right locations is giving impetus to the creation of community service jobs of various kinds. Many of these jobs may end up being paraprofessional human service jobs (Center on Budget and Policy Priorities 1997). At the same time, the immense transformation of our health care system, including Medicaid, to managed care has, for a number of reasons, increased the utility of community outreach workers (or paraprofessionals) for recruitment, disease prevention, public health functions, and chronic disease management (SEEDCO 1996; U.S. Department of Health and Human Services 1994a, 1994b; Rosenthal 1996; Harrison Institute for Public Law 1997).

The renewed contemporary interest in paraprofessionals recalls a bold War on Poverty experience of 30 years ago—the New Careers movement, one of the most thoughtful, comprehensive, and controversial efforts to launch public employment labor force planning. We begin this chapter by reviewing the assumptions, design elements, program diversity, and outcomes of the New Careers experience. We follow this historical framework with a review of paraprofessional jobs programs today, with a focus on community health outreach and education and early childhood development, and then discuss the issues facing these paraprofessional sectors, as well as mental health, family services, and youth development. We conclude the chapter with a discussion of the prospects and challenges for pursuing double social utility in the context of government cuts, welfare change, and the turn to managed care.

New Careers: Theory and Practice

New Careers advocates effectively built a movement to support paraprofessional jobs in the human services upon several strong policy currents of the late 1950s and early 1960s. An opportunity to pull these strands together into a politically marketable and multifaceted program for action occurred with the development and passage of the federal War on Poverty program in 1964 and subsequent poverty alleviation legislation. The aspirations and design tenets of the New Careers movement are best captured in the 1965 book by Arthur Pearl and Frank Riessman, *New Careers for the Poor.*

A primary policy current of the time was the perception that the velocity of automation and technological change was destroying jobs and outpacing private sector job creation. Jobs reduction of that type particularly affected minorities, many of whom had recently migrated to northern central cities. Related to this process of economic change was the understanding that, despite overall economic growth, social and economic barriers still prevented low-income, minority workers in inner-city neighborhoods and rural regions from obtaining jobs. At the same time, projections showed that human service jobs would be growing in the ensuing decades because of population growth and policy directions, in part as a response to the effects of poverty. Finally, policy responses aimed at alleviating poverty and its effects on youth and families, stretching back to settlement houses early in the century, were based on the belief that the employment and leadership development of indigenous residents was a key to changing the social structure of poor communities and transforming these communities' external relations with public systems and institutional service providers. Programs such as Mobilization for Youth on Manhattan's Lower East Side and Lincoln Hospital's mental health program in Harlem became models for hiring local people to perform valuable services as a part of community and formal system change.

The basic premise of the New Careers movement was that poor people could be hired, trained, and placed in entry-level jobs to provide valuable social services that took advantage of and relied upon their geographic, cultural, and functional similarities to low-income residents (Pearl and Riessman 1965; Cohen 1975). In addition, New Careers advocates believed that the tasks performed by paraprofessionals could be distinguished from and made compatible with those of human services professionals while bringing social services closer to the neighborhoods and residents for which they were intended. Perhaps the most important aspect of New Careers thinking from a jobs perspective was that advocates believed that entry-level training and on-the-job work experience in paraprofessional positions could, when combined with appropriate skill training, credentialing, and long-term job security, lead to career ladders into the professions. This would provide a route out of poverty for low-income workers, meet the labor demand of human services, and dramatically change how human services were delivered.

The Economic Opportunity Act of 1964 served as the initial vehicle for enabling a broader demonstration of the New Careers approach. The Scheuer Amendments to this act in 1966 provided $33 million to develop training and employment projects using paraprofessionals (Pickett 1994). Other legislation and amendments, including the Public Employment Service Act of 1970, expanded the reach of "New Careerists" during the next decade to the areas of health, education, crime control, and community beautification. In the 1970s, the Comprehensive Employment and Training Act became an additional vehicle for launching paraprofessional careers for the poor, as federal human resources policy devolved authority to states and localities. By one estimate, there were 20,000 New Careerist paraprofessionals in 1970, with 250,000 to 400,000 others working in human service agencies. By the mid-1970s, another estimate put the number of paraprofessionals at 1.5 million (Pickett 1994; Cohen 1975).

Was the New Careers movement successful? Frank Riessman, one of the early and key advocates of the training of paraprofessionals, summarized in 1984 many of the achievements of the movement. Jobs were certainly created. Low-income people with few skills received training to perform important human services. Contractual representation by unions and credentialing through community colleges helped to institutionalize the paraprofessional position as well as increase compensation and career options. Riessman concluded, however, that paraprofessionals had become absorbed by the traditional human service paradigm rather than served as change agents for human service reform.

Additional concerns about New Careers initiatives were that low-income and minority residents obtained fewer and fewer of the paraprofessional positions over time. This phenomenon of higher-skilled people displacing low-income, lesser-skilled people was in part a function of the economic downturn and climbing unemployment rates of the 1970s. The labor market became more competitive. In general, the important career ladder component of the New Careers vision performed badly as well—paraprofessionals were not able to sustain participation in these activities even in the few cases where they were available (Cohen 1975).

Paraprofessional Job Creation Today

Since Riessman's retrospective look at New Careers in the mid-1980s, there has been renewed interest in the deployment of paraprofessionals as a strategy of both good practice and economic development, if not full-scale reform. Our definition of *paraprofessional* is similar to the one used by the New Careers movement:

> a staff member working in the field of human services who: may or may not have a formal academic degree in the field in which s/he is working, but has received specific training to do so; works semi-autonomously to provide direct services to clients; receives a salary and possibly other benefits to compensate for the work performed; receives direction and supervision from a professional colleague; and is likely to be indigenous to the population served. (Wong 1995)

In the 30 years since New Careers, the number of paraprofessionals hired in the health and human services fields has grown steadily, though the pace is likely to slow with contemporary reductions in government services (U.S. Department of Labor 1994a, 1994b). The ways in which paraprofessionals are deployed are likely to continue expanding, however, because of both government cost cutting, which contributes to the displacement of professionals, and the efforts of reformers to come up with new responses to intractable social problems. Table 9.1 lists 75 job titles and job functions that are used to describe paraprofessional work in the fields of health, education, family services, mental health, and juvenile justice.

There is a fugitive quality to the data on paraprofessionals, and that feature, combined with numerous variable job titles and descriptors, makes it nearly impossible to count the number of human service paraprofessionals in the workforce today (Wong 1995). Many paraprofessional titles are not recognized

TABLE 9.1 Descriptors and Titles Commonly Used to Identify Paraprofessional Personnel in the Health and Human Services

Adjunct therapist	Family health promoter	Natural caregiver
Aide	Family preservationist	Natural helper
Allied human service personnel	Family resource person	Neighborhood adviser
	Family service coordinator	Neighborhood advocate
Assistant	Health facilitator	Nonprofessional
Auxiliary personnel	Health liaison	Nutrition assistant
Caseworker assistant	Health program aide	Outreach worker
Child development aide	Health promoter	Paraeducator
Child development associate	Health visitor	Paratransit operator
	Home aide	Parent aide
Child development worker	Home care aide	Parent befriender
Classroom aide	Home health aide	Parent helper
Client advocate	Home visitor	Peer counselor
Community health adviser	Indigenous health aide	Peer health adviser
Community health advocate	Indigenous helper	Peer health educator
	Indigenous paraprofessional	Peer helper
Community health aide	Informal helper	Preprofessional
Community health outreach worker	Instructional aide	Promoter
	Instructional associate	Psychiatric aide
Community health representative	Lay community health worker	Resource mother
		Semiprofessional
Community health worker	Lay family support person	Social work aide
Community helper	Lay health adviser	Subprofessional
Family advocate	Lay health worker	Teacher aide
Family aide	Lay helper	Teacher associate
Family development aide	Lay volunteer	Tracker
Family development worker	Mental health technician	Volunteer health educator

SOURCE: Wong (1995).
NOTE: Not all titles are recognized in U.S. Department of Labor literature, nor are all titles in the DOL literature listed here.

in the U.S. Department of Labor literature, and many DOL titles combine professional and paraprofessionals. Reliable estimates of the number of paraprofessionals working in schools converge at 500,000 (Pickett 1995a). It is not unreasonable to guess that another 200,000 paraprofessionals are working in health care, child care, mental health, family support services, and juvenile justice.

Although there are limited data on absolute numbers of paraprofessional workers, a review of the human service literature suggests consensus on the key roles of paraprofessionals. Across the fields of health, education and child care, family and children's services, mental health, and juvenile justice, paraprofessional work encompasses seven key roles:

- *Outreach:* identifying and engaging hard-to-reach clients
- *Education and training:* teaching skills to adults and children
- *Emotional support:* responding with empathy to often isolated clients

EXAMPLES

- *Mentoring and role modeling:* using self as an example of desired behavior
- *Concrete services:* brokering services from formal service systems, providing or securing transportation, providing housing assistance, providing help shopping for and preparing food, and so on
- *Surveillance:* observing client compliance with medical regimen or other behavioral goals
- *Clerical:* record keeping, case documentation, financial reporting

Paraprofessionals in Health

Perhaps the most recognizable paraprofessionals in health care are nurse's aides and home care workers. Steven Dawson has addressed in Chapter 7 of this volume how an innovative worker ownership model has been used to improve jobs and career opportunities in the home health field. For this reason, we focus here on the people who make up another large sector of the paraprofessional health care workforce—those who operate as community outreach workers.

Known as *community health outreach workers, community health advisers,* and by many other titles, these workers are lay staff hired more often for their knowledge of and credibility in particular target communities than for their particular substantive expertise (U.S. Department of Health and Human Services 1994a, 1994b). Community health advisers, or CHAs (the term we will use to refer to all community health outreach staff), combine all of the core activities described above, with an emphasis on education, emotional support, modeling, and service brokering. CHAs tend to be employed in the low-income urban and rural communities in which they live, and many of their assignments focus on linking poor families to formal services and supporting these families in their efforts to cope with poverty and its stresses. Although this role has antecedents in the 19th century's wealthy "friendly visitors" and in the public health nurses affiliated with settlement houses in the early part of the 20th century, it is only since the time of the New Careers movement that lay residents of target communities have served in this role (Larner and Halpern 1987).

- Prevalence of Community Health Advisers

More than 600 programs employ approximately 13,000 CHAs in a variety of health programs (Rosenthal 1997). A recent Centers for Disease Control survey of CHA programs found that about 60% of CHAs are involved in mental health and nutrition services; half report addressing substance abuse, HIV/AIDS, and violence prevention; and one-third report focusing on chronic disease management and injury prevention (U.S. Department of Health and Human Services 1994a, 1994b). The overlapping categories used by the survey respondents reflect the flexible, interdisciplinary nature of CHA work.[1] Although CHAs often are expected to know a little about a lot of fields, most get minimal training, work fewer than 10 hours a week, and are paid $5.00 to $9.00 per hour (Brownstein 1996). In fact, only about 77% of CHAs are paid for their services (Zielenbach 1995). The fact that volunteers and paid paraprofessionals perform much the same work reflects a debate in this field about whether wages compromise CHAs' accountability to their communities. This belief in the inherent value of unpaid CHA work is a substantial, albeit minority, opinion in the field (Israel and Rounds n.d.; Eng and Young 1992). We do not focus on it here, in large part because unpaid CHAs tend to be middle-class and have other jobs or do not need to work (Eng and Smith 1995).

Many CHAs find employment directly or indirectly through three federally funded programs: the Indian Health Service, the Cooperative Extension System, and Healthy Start. The Community Health Representative Program of the Indian Health Service, which employs 1,500 CHAs, is the only CHA program directly funded and authorized by the federal government (Harrison Institute for Public Law 1997). The Indian Health Service program differs from most CHA programs in that staff members are paid better, about $11.00 an hour, and some hold professional degrees. The U.S. Department of Agriculture's Cooperative Extension Service operates the Expanded Food and Nutrition Education Program, which utilizes 3,000 part-time paraprofessionals and 50,000 volunteers at schools, work sites, and other locations to provide nutritional education and teach basic budgeting and parenting skills (U.S. Department of Health and Human Services 1994b). Healthy Start, a 6-year program whose first round of federal funding ended in September 1997, is a 15-city demonstration with the ambitious goal of reducing infant mortality by 50%. Most of the Healthy Start sites hired indigenous paraprofessionals, frequently former welfare recipients, for part-time CHA jobs that pay an average of $6.00 an hour (McDermott 1995; Squires 1997; Grier 1995).

■ Training

CHAs usually receive about 40 hours of training from a nurse or social worker before starting their jobs, followed by in-service training every 2 to 4 months on general disease prevention, communication skills, and community resources. Although about half of the programs hiring CHAs offer certificates, only about 10% of the available training is eligible for college credit (Zielenbach 1995).

There are a handful of model CHA training programs with track records of preparing well-qualified CHAs, issuing credentials recognized by public agencies and other major employers, and exhibiting high placement rates, but these are the exception (SEEDCO 1996; U.S. Department of Health and Human Services 1994b). Although quality training for CHAs and their supervisors is considered one of the most important factors in program effectiveness, the supply is inadequate, and most of what does exist is unevenly executed and poorly funded (Whitebrook 1993). Career ladders have been similarly limited, although the federal government and foundations have been paying increased attention to this issue (Harrison Institute for Public Law 1997; Rosenthal 1996).

■ Effectiveness

Although comprehensive data are limited to only a few programs, CHAs are widely regarded as improving the quality of health care in low-income neighborhoods. Data from Baltimore's Healthy Start program and a similar model in Oregon suggest that CHAs are successful in reducing the incidence of premature births and low-birth-weight babies, as well as reducing health care costs by decreasing use of emergency rooms and the incidence of births requiring special care (Baltimore City Healthy Start 1997; Healthy Start n.d.; Poland et al. 1991). Findings on effectiveness and cost savings have also been documented in programs using CHAs for chronic disease management (Chaulk et al. 1995; Moore et al. 1996). CHAs are generally considered to be more readily accepted by families and community members than are professionals, and for this reason are found to be more effective in eliciting information from clients, allaying fears about professional medical help, conveying information credibly, and providing social support (Larner and Halpern 1987; Wasik and Roberts 1994; Watkins et al. 1994).

■ Barriers to Growth

Many of the impediments to wider use of CHAs or the improvement of CHA jobs are related to systemic issues of limited training and credentialing and virtually no cross-sector recognition of either. Career ladders are currently nonexistent, though their potential is gaining the attention of policy makers. It seems that when CHAs move up, it is often out of their fields. Some Healthy Start programs were able to help CHAs find jobs paying $7.50 per hour in marketing, office management, and technology by focusing on the development of particular related skills (Grier 1995). Other barriers have more to do with conflicts inherent in the job. Although community know-how is frequently cited as a factor in CHAs' success, some clients do not want neighbors knowing the details of their family lives, and others dismiss the capacities of lay workers. Some difficulties stem from the life conditions of CHAs themselves. Hired because they share some of the experiences of their clients, CHAs often share their clients' problems as well, which challenges their ability to work. Another key impediment is the resistance of CHAs' family members who feel threatened by the increased independence that comes with work.

■ CHAs and Managed Care

Managed-care companies are insurers that enter into contracts that allocate or share risk to provide funding to pay care providers, hospitals, or other parts of the health care service system. Managed-care arrangements have grown rapidly over the past 10 years for all U.S. families and are becoming the predominant form of health care for low-income families receiving Medicaid. Half of all families receiving Medicaid nationally are enrolled in managed-care plans; in some states the proportion is as high as 75% (SEEDCO 1996). By the end of 1997, 31 states had mandatory Medicaid managed-care programs.[2]

Managed-care companies, particularly those serving Medicaid clients, are trying to establish relationships with CHA programs because CHA outreach and education appear to be cost-effective in reducing emergency care among their Medicaid clients. Emergency room care and subsequent hospitalization represent 40-60% of medical costs for managed-care companies, and Medicaid clients are three times more likely than other patients to use emergency rooms and to be hospitalized as a result (SEEDCO 1996). Low-birth-weight and premature babies, though only 7% of all births, occur disproportionately among Medicaid patients and consume 60% of all health funds for newborns (Harrison Institute for Public Law 1997).

Many CHA agencies are interested in entering into contracts with managed-care companies, particularly to replace revenue lost by the winding down of federal programs such as Healthy Start. Many CHA agencies currently receive income from managed-care companies, but most contracts are small, experimental, and more akin to grants. Barriers to expansion of these relationships can be found on both sides. Managed-care companies are hesitant to engage with CHA providers that cannot document their outcomes or cost savings, or clearly define the services they provide and their prices (SEEDCO 1996; Orbovich 1996). CHA agencies question the market mentality of managed-care companies and are wary that their outreach services will be reduced to enrolling new plan members.

Some CHA programs are working to adapt their administrative and managerial capacities to meet the requirements of managed-care companies. Baltimore's Healthy Start program, for example, has packaged data from its home visiting program to show how it reaches managed-care goals of improved birth outcomes, reduced emergency care, and increased patient compliance (Baltimore City Healthy Start n.d.). Baltimore's and other Healthy Start programs are negotiating contracts with managed-care companies to provide their full complement of CHA services (Squires 1997).

As an attempt to position CHA services better in a managed-care market, SEEDCO, an economic development intermediary, is developing a model CHA agency that seeks to expand employment for CHAs by designing new services, contracting with managed-care companies, and documenting cost and health benefits of their program (SEEDCO 1996). The experiment builds on the work of Cooperative Home Care Associates (CHCA) in the Bronx, described in Chapter 7 of this volume. SEEDCO is exploring whether a CHA intermediary organization can be formed and sustained from managed-care revenues and public health dollars that will more efficiently organize CHAs; provide a focus on CHA training, compensation, and careers; and ensure public health functions. Because CHAs are an emerging job category and Medicaid managed care is in transition in many states, SEEDCO's strategy is to use this time as an opportunity to affect the definition of CHA work, establish career paths, and influence job structures, wages, and benefits.

Paraprofessionals in Education and Child Care

Paraprofessionals working in schools and child-care centers form perhaps the single largest group of paraprofessionals working today, numbering roughly 600,000 full-time workers. They also have the most institutionalized jobs, with continuous if not stable federal funding, recognized career ladders in education, and an infrastructure of training and credentialing programs. They are also working in a growth industry. The number of paraprofessionals in schools has grown by 100,000 in the past 5 years, and there is a national shortage of bilingual teachers (Pickett 1995b). An adequate supply of child care is a linchpin for successful welfare-to-work efforts under the new federal law, and there is growing interest in expanding the number and capacity of child-care centers for working parents (National Center for the Early Childhood Workforce 1996a; "Working and Welfare Parents" 1997).

Much of the relative stability of paraprofessional work in education and child care has its foundation in the federal legislation of the 1960s and 1970s that established Head Start and Title I early childhood development and education programs. These programs made federal funds available to schools and community organizations specifically for the hiring of paraprofessionals. With passage of the Individuals With Disabilities Act in 1974, paraprofessionals increasingly assured the availability of individualized programs for students and preschoolers with disabilities (Pickett 1995c). Demand for Head Start teachers also increased when the program expanded to include low-income or disabled 4- and 5-year-olds.

■ Training

Training for paraprofessional work in schools and child-care centers, though different in structure and content, shares similar features but also illustrates the frustrations of institutionalizing decent-paying jobs and standards of practice recognized within and across human service systems. Both education and child care support a network of training programs that produce graduates shown to deliver better service, stay in their jobs longer, and fill specific skills shortages in their field. However, in education there is no consensus on the skills paraprofessionals should have, nor is there any standardized set of training or professional development opportunities (Pickett 1995a). In large part because of collective bargaining, paraprofessionals in education have some limited career ladder opportunities and enjoy relatively higher wages (Ginsburg 1996; Gould 1996; Brandick 1994; Pickett 1997). In child care, the federal government has established and supports a credential for paraprofessional child-care workers that is often cited as a model for other fields; however, many states do not require credentialing for child-care workers, and wages in the field are at the poverty level (Whitebrook, Phillips, and Howes 1989).

■ School Staff

There are almost 150 college- and university-based training programs nationally that enroll more than 9,000 paraprofessionals. Participants in these programs meet workforce needs in several ways. First, more than three-fourths are minority members, compared with 13% of teachers and one-third of all students.

Second, they specialize in fields in which there are labor shortages, such as special education, bilingual programs, and urban school systems. Finally, they stay in their jobs longer and report relatively low attrition rates, averaging 7% (Recruiting New Teachers 1996).

These trainees are a small minority of the paraprofessionals entering school-based jobs. Of all new paraprofessional hires, 70-90% have no prior training (Pickett 1994). Both trained and untrained new hires enter a system independent of any comprehensive system of career development. Most school systems have no in-service or on-the-job training for paraprofessionals and no connection between postsecondary education for paraprofessionals and the credentials required for a professional teaching job. As a consequence, paraprofessionals have limited access to the one million new teacher jobs projected for the next 5 years. At the same time, school districts hire 50,000 unlicensed teachers each year, many without any recent classroom experience or any such experience at all (Recruiting New Teachers 1996). Although it is hard for paraprofessionals in education to make the jump to teaching jobs, 11 states have established criteria for hiring, training, and advancement within the paraprofessional ranks (Pickett 1994).

■ Early Childhood Development

The overwhelming majority of early childhood development teachers are hired with no previous training, and they have limited opportunities for in-service or on-the-job training (Copple 1991; Whitebrook, Phillips, and Howes 1989). There are, however, important exceptions, particularly in the area of continuing professional development. During the peak of the New Careers movement, the Head Start program initiated a national training and credentialing program for early childhood workers that it continues to support. The child development associate (CDA) credential qualifies trainees to teach in Head Start programs and is recognized by 49 states in their licensing requirements for early childhood providers. Part of Head Start's emphasis on hiring and promoting current and former Head Start parents, who compose more than a third of Head Start staff, CDA training is accessible to workers who might not be able to enroll in college-level early childhood education programs. About 45% of Head Start's staff have the CDA credential or its equivalent, and every Head Start classroom must have at least one such certified teacher. CDA-qualified Head Start teachers stay in their jobs much longer than their peers, are more likely to be promoted, and contribute to improved outcomes for their students (Henry 1996; Howes and

Galinsky 1995). However, their promotions rarely translate into significant increases in income, with most raises under $500 a year (Henry 1996). In a field where the average entry-level salary is less than $10,000 and wages after 6 to 10 years of teaching are commonly $11,000, the CDA-related wage does not go far (National Center for the Early Childhood Workforce 1996b). Data also indicate that workers holding CDAs are more likely to be already better educated than other child-care workers and to be white, suggesting that the benefits of the CDA are not evenly distributed among the child-care workforce (Henry 1996; National Center for the Early Childhood Workforce 1997b).

■ Effectiveness

The literature on the relative effectiveness of paraprofessionals in education is limited, and what exists in the early childhood development field mostly provides comparisons between credentialed and less-trained paraprofessionals. A recent assessment of the nation's leading accreditation system for child-care centers found that centers most likely to provide high-quality care are those that pay higher-than-average salaries as well as receive accreditation (National Center for the Early Childhood Workforce 1997b). Other analyses support the conclusion that key indicators of quality care are the combined factors of higher wages, more training for workers, and more staff per child, as well as accreditation, higher overall program spending, and funding sources other than parent fees (Helburn et al. 1995). In education, where there is virtually no documentation of outcomes, many consider paraprofessionals to be a key to rethinking schools as community-based institutions, because they are often the only school staff members who live in the same neighborhoods as the students. In addition to their classroom roles, these paraprofessionals often function as combined security force, guidance counselors, academic prods, and moral compasses for children (Cipollone 1995).

In early childhood development, studies of CDA-credentialed staff suggest that they are more likely to contribute to improved outcomes for children and to stay longer in their jobs, and, because of their longer tenure, the enhanced training is more cost-efficient than continually orienting new staff (Howes and Galinsky 1995; Whitebrook, Phillips, and Howes 1989). Strategies for reducing turnover are key, given the link between positive child outcomes and a stable workforce, and the fact that one-third of all child-care teachers leave their centers each year (National Center for the Early Childhood Workforce 1994, 1997b). Head Start and other federal funding programs offer limited incentives to states to provide career ladders for experienced early education teachers. The steps to a senior teacher title do not translate into significant increases in wages.

■ Barriers

The challenges for paraprofessionals in education and early childhood education have less to do with increasing the number of jobs for low-income residents—that seems to be happening already—than with improving the quality of available jobs. In education, unions have developed some career ladder programs that lead to teaching positions, and collective bargaining has brought relatively higher wages for paraprofessionals in schools. However, the career ladder programs operate on a small scale and only in sectors experiencing labor shortages. Even in markets with labor shortages, a more common way to fill teacher slots is with paraprofessionals working at lower title and salary. Nationally, 20% of teacher's aides funded by federal compensatory education programs (Title I) instruct students without any teacher supervision. In California, 40% of bilingual instructors are aides (Recruiting New Teachers 1996).

Using paraprofessionals in schools as direct replacements for teachers exacerbates tensions between the two levels of workers. This competition combines with three other factors to hinder the upgrading of paraprofessionals' jobs in schools and their upward mobility: a low rate of transfer by education students from community colleges to senior colleges, no uniformity of standards across states, and a system for financing community colleges that provides more funds for technical training programs than for those in human services (Pickett 1995c). Recent shifts in federal policy, however, suggest a promising new direction. The 1997 reauthorization of the Individuals With Disabilities Act calls on states and locales to ensure that paraprofessionals in schools receive appropriate training and supervision. Although the provisions do not require states to establish standards of practice or training, the legislation helps bolster the efforts of those working for policy changes locally.

The early childhood development field has made strides in credentialing and raising the quality of paraprofessional work, as well as expanding job access to low-income people. Once hired and trained, however, workers stay poor. The low wages of child-care teachers have multiple systemic causes (Helburn et al. 1995; National Center for the Early Childhood Workforce 1996a):

1. States have minimal licensing requirements, and low-wage, unregulated care depresses overall wages.

2. Many child-care centers serve parents with limited abilities to pay and receive only partial reimbursement from the state.
3. There has been growth among private, for-profit child-care centers, which pay lower wages than nonprofits.
4. Voucher systems have expanded, leading to volatile enrollment at centers and consequent destabilization of employment and training.
5. Parents have difficulty in observing the quality of care at centers, which means limited demand for higher-quality care.

As noted above, low wages adversely affect not only staff members at child-care centers; because of the effect of staff turnover on child outcomes, they also hurt the children in the centers.

There have been some activities taking place in early childhood development that indirectly address low compensation for teachers, and recent efforts have focused specifically on wages and working conditions. A range of financing programs for nonprofit child-care facilities are funding the improvement, expansion, or start-up of child-care centers (Center for Policy Alternatives 1996; North Carolina Facilities Fund n.d.). State loans and loan guarantees, the work of financial intermediaries, and microlending programs for child-care centers can indirectly improve the quantity and quality of jobs for child-care teachers. Activities in this area are expected to receive a boost because the 1996 Community Reinvestment Act regulations have been amended to include child care as a viable community lending option (Center for Policy Alternatives 1996). Childspace, Inc., is a worker-owned child-care network in Philadelphia, now expanding to several other regions, that combines child-care provider ownership and participation with investments in training and career ladders (Pitegoff 1993).

More recent efforts have focused on the development of a variety of employee associations to negotiate improved wages and working conditions for child-care teachers (National Center for the Early Childhood Workforce 1996b). Unlike school paraprofessionals, who are covered by collective bargaining in 35 states, only 4% of child-care teachers belong to unions (AFL-CIO Public Employee Department 1995; National Center for the Early Childhood Workforce 1996b). Although unionized child-care programs have higher wages and lower turnover, the typically small staff size at child-care centers, high employee turnover, and limited funds available for improved compensation make child care a challenging field in which to organize collective bargaining units (National Center for the Early Childhood Workforce 1997a). There are signs that this may be changing, however. Several labor unions, including the American Federation of

State, County and Municipal Employees, the American Federation of Teachers, the National Education Association, Service Employees International Union, and the United Auto Workers, have shown a growing interest in representing lower-wage service sector employees.

Other recent efforts to raise the wages of child-care workers have been undertaken in Massachusetts and North Carolina, both of which have increased reimbursement rates specifically for worker wages and provide additional compensation to workers who complete additional training (National Center for the Early Childhood Workforce 1997a). The federal Head Start program has also allocated additional funds to boost wages of staff by an average of $1,500 per year. States also have access to funds through the federal Child Care and Development Block Grant designated for quality improvements. This quality set-aside has resulted in initiatives in training, licensing, and compensation in more than a third of states (Azer, Caparo, and Elliott 1996).

Cross-Sector Policy Issues

Although paraprofessional services in health care, education, family support, and juvenile justice developed largely independent of each other and continue to operate separately, the policy issues they raise are similar. Family support is a relatively new field, built upon a New Careers hypothesis that paraprofessionals can provide effective bridges for disadvantaged communities by drawing upon informal support systems—kin, neighbors, self-help groups—to aid family coping and promote well-being (Halpern 1986; Family Resource Coalition 1996; Dawson et al. 1991; Weiss and Jacobs 1988; U.S. General Accounting Office 1997; National Economic Development and Law Center 1996). Mental health paraprofessional positions also derive, in part, from the New Careers experience, but their history is rooted in the emergence of self-help and mutual-aid groups in the late 1970s as well (Van Tosh et al. 1993; Wells, Anderson, and Popper 1993). Although there have been a number of experiments in the use of paraprofessionals in the juvenile justice arena since the 1960s, paraprofessional workers have been less well received there than in other sectors (Allen et al. 1985; Borduin 1994; Gordon and Arbuthrot 1988; Krisberg and Austin 1993). Today, paraprofessionals in the juvenile justice system, often called *trackers,* assume roles not unlike CHAs or family support and mental health workers, although surveillance is a higher priority.

Paraprofessional utilization and career ladder expansion in all these human service sectors face four crosscutting challenges, which we discuss below.

Lack of Clarity About Paraprofessional Roles

In all human service sectors where paraprofessionals are deployed, ambiguity exists about the responsibilities of paraprofessionals. Professionals and paraprofessionals are deployed interchangeably in the same jobs. In some cases, as in education, this can be part of an effort to displace professionals or to cope with staff shortages; in other cases, such as in mental health, it is more symptomatic of the overly generalized job descriptions often used for paraprofessional work.

Another reason human service systems are unclear about paraprofessional roles is that those sectors are uncertain about the jobs professionals should do. It is by no means clear to what extent professional training and credentials are necessary for individuals to perform particular health, education, and social service functions (Kinney et al. 1997). This is particularly notable in the evaluation literature for mental health services, which suggests that in some cases, there is no difference between professional and paraprofessional care, or in CHA programs, where similar interventions variously deploy nurses, paraprofessionals, and volunteers (Hattie, Sharply, and Rogers 1989; Aiken et al. 1989; Koroloff et al. 1994).

Poorly defined roles engender tension between professionals and paraprofessionals, not only because professionals fear displacement, but also because neither understands fully what the other is capable of doing. Professionals are reluctant to risk their liability by giving paraprofessionals too much authority in program decisions, and paraprofessionals do not learn when to turn to professionals for guidance. As a consequence, paraprofessionals do not learn either their limits or their potential; they are locked out of opportunities for advancement. Professionals, for their part, lose opportunities to delegate responsibility that would free them for tasks that truly require professional skills.

Fragmented Training, Credentialing, and Standards of Practice

Community health outreach, school-based and early childhood education, and family support all have model training programs, but they are few and far between. Most agencies develop their own training and practice guidelines, which may or may not compare with those of other programs in the same field. With the exception of training for the CDA credential, which qualifies individuals for some school-based positions, no training is recognized by more than one sector.

Even when there are proven training approaches, or national or regional credentials, as in the case of early childhood education and family support, their success in improving the quality of jobs and care depends on state licensing and financial aid policies. If states do not require credentials, subsidize regulated child care, or reimburse school tuition, standards of training or practice will not become prevailing policy.

Without a hierarchy of core knowledge and skills, it is impossible to describe a pathway that can lead paraprofessionals to higher-level jobs. To the extent that career ladders exist in human services for paraprofessionals, they tend to be levels *within* paraprofessional occupations, not *between* paraprofessional and professional jobs, and they tend to be in those fields in education and early childhood most influenced by statutory federal funding.

There is also the risk that more extensive credentialing structures will inflate the formal skills necessary for paraprofessional jobs and limit access to these jobs for low-income residents with primarily indigenous skills. Paraprofessional jobs in education, for example, attracted greater numbers of middle-class workers in the 1970s and 1980s as the jobs became more structured and better compensated. Similarly, as we have noted, child-care workers with CDA credentials are more likely to be better educated and to be white than are other child-care workers. The entry point for paraprofessional jobs tends to shift as the jobs improve, unless there are deliberate, systemic efforts to preserve a community orientation. The Indian Health Service, for example, funds some of the highest-quality community health adviser jobs, but because of an institutional priority placed on community know-how, these jobs are still accessible to residents. College- and university-based training programs for paraprofessionals in education often specialize in bilingual and urban education, and consequently train higher numbers of indigenous paraprofessionals.

Uneven and Inadequate Supervision

Supervision is the most overlooked core element of successful paraprofessional work. Paraprofessionals with scant prior work experience and little more than a high school equivalency diploma, employed in largely unstructured human service jobs, are expected to plan the use of their time for meetings with clients and keeping records. Their orientation training gives them a limited repertoire of observation and assessment skills that they will need to apply to a wide variety of client situations. They are isolated, working away from any "central office," often under great stress, and at some risk to their personal safety.

The task of supervising such workers is more educational than administrative. An evaluation of family support programs funded by the Ford Foundation found that supervisors' success depended on their ability to motivate, teach, nurture, and evaluate lay workers (Larner, Halpern, and Harkavy 1992b). Yet the limitations of paraprofessional training programs are exceeded only by those of their supervisors. Many professionals have never worked with paraprofessionals before receiving the responsibility of supervising such workers, and may be ignorant of their needs and ambivalent about their value.

Low Wages

One thing that CHAs, community mental health workers, teacher's aides, trackers, and family support paraprofessionals have in common is low wages. For many human service paraprofessionals, the issue is not whether wages are family supporting, but whether those wages will endanger their eligibility for public assistance or put them in competition with welfare recipients required to find work (Milgram 1995; Holzer 1996; Mishel and Schmitt 1996; "Welfare Recipients" 1997). Low wages predominate where there are established training and credentials, as in child care, and where there are labor shortages, as in education. The only factor associated with improved compensation is the presence of a union or other collective bargaining arrangement, as is the case in several state education systems (Bluestone 1995).

Conclusion

More than 30 years ago, federal legislation provided the impetus for expansion and institutionalization of a New Careers approach to human services. Today, federal—and state—policy remains a key factor, but less in direct job creation and more in reimbursement and regulation standards that set requirements for funding and practice. Recognition of this government oversight role in health care, child care, and other human services as a de facto jobs policy for paraprofessionals is the best opportunity for a new New Careers initiative.

How can government policy foster more and better paraprofessional jobs? There are two primary levers government can use. The first is the definition of reimbursable services or allowable activities under regulation or statute. In health care, for example, the states of Ohio and Wisconsin are negotiating contracts with managed-care organizations that require community health outreach as a reimbursable service in Medicaid-funded maternal and child health programs. In child care, the federal reauthorization of the Community Reinvestment Act specifies the construction of child-care centers as an eligible activity. In each case, government policy defines a new market niche for paraprofessional services.

Government policy also influences who can be hired for paraprofessional jobs that are publicly funded or regulated, as well as how those workers are to be trained and supervised. We discussed earlier how federal Head Start legislation has been used to encourage the employment of parents in child-care programs, to specify standards of practice, and to provide incentives for investment in training. In the case of training, some states have followed suit by offering additional and substantial financial support for staff development, with Massachusetts and North Carolina the most notable examples. Although the federal government has a far more limited role in education, the 1997 Individuals With Disabilities Act represents a first step in requiring, albeit without enforcement, that states provide appropriate training and supervision of paraprofessionals in schools.

In most instances, state and federal policies defining the design and delivery of human services lack a jobs focus. Above are a few examples in the fields of child care, education, and community health care of how more strategic regulation and reimbursement policies can help build a structure for better paraprofessional jobs that are still accessible to community residents. Other human services areas, such as mental health and child welfare, present other opportunities, because these sectors offer similar roles for paraprofessional work and involve significant government oversight. Considering that government policy now affects approximately 700,000 paraprofessional jobs in human services, this strategy could make a difference on a scale that has not been matched since the New Careers initiative.

Notes

1. Home visiting is a commonly used strategy in CHA tasks that has been studied extensively (see Larner, Halpern, and Harkavy 1992a; David and Lucille Packard Foundation 1993).

2. For a discussion of the impact of managed care in other human service fields deploying paraprofessionals, see Scallett, Brach, and Steel (1997).

References

AFL-CIO Public Employee Department. 1995. *Public employees bargain for excellence: A compendium of state public sector labor relations laws.* Washington, DC: AFL-CIO.

Aiken, L. S., L. A. Lo Sciuto, M. A. Ausetts, and B. S. Brown. 1989. Paraprofessional versus professional drug counselors: The progress of clients in drug treatment. *International Journal of the Addictions* 19:383-401.

Allen, H. E., C. W. Eskridge, E. J. Laksoa, and G. V. Vito. 1985. *Probation and parole in America.* New York: Free Press.

Azer, S. L., K. L. Caparo, and K. A. Elliott. 1996. *Working toward making a career of it: A profile of career development initiatives in 1996.* Boston: Center for Career Development in Early Care and Education, Wheelock College.

Baltimore City Healthy Start. 1997. Early outcome evaluation results. Unpublished manuscript, February 7.

Baltimore City Healthy Start. N.d. Handout.

Bluestone, B. 1995. The inequality express. *American Prospect* 20:81-93.

Borduin, C. M. 1994. Innovative models of treatment and service delivery in the juvenile justice system. *Journal of Clinical Child Psychology* 23:79-125 (suppl.).

Brandick, S. 1994. The Los Angeles Unified School District and the Service Employees International Union Local 99 paraeducator career ladder. *Newsletter of the National Resource Center for Paraprofessionals in Education and Related Services* 17(Spring):2-9.

Brownstein, N. 1996. Personal communication. Centers for Disease Control and Prevention, Atlanta, GA, March 15.

Bruner, C. 1996. Personal communication. Child and Family Policy Center, Des Moines, IA, October.

Bush, M., A. M. Ortiz, and A. Maxwell. 1995. *Tracking the dollars: State social service spending in one low-income community.* Chicago: Woodstock Institute.

Center for Policy Alternatives. 1996. *Financing the future: Innovations in child care financing.* Washington, DC: Center for Policy Alternatives.

Center on Budget and Policy Priorities. 1997. *Enhancing income security through public job creation.* Washington, DC: Center on Budget and Policy Priorities.

Chaulk, C. P., K. M. Rice, R. Rizz, and R. E. Chaisson. 1995. Eleven years of community based directly observed therapy for tuberculosis. *Journal of the American Medical Association* 274:945-951.

Cipollone, A. 1995. Personal communication. Annie E. Casey Foundation, Baltimore, March 2.

Cohen, R. 1975. *"New Careers" grows older: A perspective on the paraprofessional experience 1965-75.* Baltimore: Johns Hopkins University Press.

Copple, C. 1991. *Quality matters: Improving the professional development of the early childhood workforce.* Washington, DC: National Institute for Early Childhood Professional Development.

David and Lucille Packard Foundation. 1993. The future of children. *Home Visiting* 3(3).

Dawson, P. M., J. L. Robinson, P. M. Butterfield, W. T. Van Doornick, T. J. Gaensbaver, and R. J. Harmon. 1991. Supporting new parents through home visits: Effects on mother-infant interactions. *Topics in Early Childhood Special Education* 10(4):29-44.

Eng, E., and J. Smith. 1995. Natural helping functions of lay health advisors in breast cancer education. *Journal of Breast Cancer Research and Treatment* 35(March):23-29.

Eng, E., and R. Young. 1992. Lay health advisors as community change agents. *Family and Community Health* 15(1):29-40.

Family Resource Coalition. 1996. *Guidelines for family support practice.* Chicago: Family Resource Coalition.

Ginsburg, L. 1996. Personal communication. AFL-CIO Public Employee Department, March 26.

Gordon, D. A., and J. Arbuthrot. 1988. The use of paraprofessionals to deliver home-based family therapy to juvenile delinquents. *Criminal Justice and Behavior* 15:366-378.

Gould, J. 1996. Personal communication. American Federation of Teachers, Washington, DC, February 6.

Grier, E. 1995. Personal communication. Maternity and Infant Health Care, Cleveland, OH, January 13.

Halpern, R. 1986. Community based support for high risk families. *Social Policy* 11(Summer):47-50.

Harrison Institute for Public Law. 1997. *Community health worker programs: A leadership brief on community preventive health.* Washington, DC: Georgetown University Press.

Hattie, J. A., L. F. Sharply, and H. J. Rogers. 1989. Comparative effectiveness of professional and paraprofessional helpers. *Psychological Bulletin* 95:539-591.

Healthy Start, Inc. N.d. *Healthy Start overview.* Portland, OR: Healthy Start, Inc.

Helburn, S., C. S. Howes, D. Bryant, S. L. Kagan, M. Culkin, J. Morris, N. Mocan, L. Phillipsen, R. Clifford, D. Cryer, E. Peisner-Feinberg, M. Burchinal, and J. Rustici. 1995. *Cost, quality, and child outcomes in child care centers: Public report.* Denver: Economics Department, University of Colorado.

Henry, M., ed. 1996. *The 1994 survey of CDA's: A research report.* Washington, DC: Council for Early Childhood Professional Recognition.

Holzer, H. J. 1996. *What employers want: Job prospects for less-educated workers.* New York: Russell Sage Foundation.

Howes, C. S., and E. Galinsky. 1995. *The Florida Child Care Quality Improvement Study.* Interim report. New York: Families and Work Institute.

Israel, B., and K. Rounds. N.d. Social networks and social support: A synthesis for health educators. *Advances in Health Education and Promotion* 2:311-351.

Kinney, J., K. Strand, M. Hagerup, and C. Bruner. 1994. *Beyond the buzzwords: Key principles of effective frontline practice.* Working paper. Des Moines, IA: National Center for Services Integration.

Kinney, J., K. Apple, S. Berstein, K. Fogg, L. Fogg, D. Haapala, E. Johnson, R. Johnson, J. Nittoli, D. Price, K. Roberts, T. Steele, K. Robinson, E. Trent, M. Trent, V. Trent, R. Smith, and R. Vignec. 1997. *Walking our talk in the neighborhoods: Going beyond lip service in service delivery improvement.* Working paper, draft, February 11.

Koroloff, N. M., D. J. Elliot, P. E. Koren, and B. J. Friesen. 1994. Connecting low-income families to mental health services: The role of the family associate. *Journal of Emotional and Behavioral Disorders* 2:240-246.

Krisberg, B., and J. F. Austin. 1993. *Reinventing juvenile justice.* Newbury Park, CA: Sage.

Larner, M., and R. Halpern. 1987. Lay home visiting programs: Strengths, tensions and challenges. *Bulletin of the National Center for Clinical Infant Programs* 3:179-197.

Larner, M., R. Halpern, and O. Harkavy, eds. 1992a. *Fair Start for Children: Lessons learned from seven demonstration projects.* New Haven, CT: Yale University Press.

———. 1992b. The Fair Start story: An overview. In *Fair Start for Children: Lessons learned from seven demonstration projects,* edited by M. Larner, R. Halpern, and O. Harkavy. New Haven, CT: Yale University Press.

McDermott, M. 1995. Personal communication. Case Western Reserve University, Cleveland, OH, January 13.

Milgram, D. 1995. Personal communication. New Traditions for Women, Washington, DC, March 6.

Mishel, L., and J. Schmitt. 1996. *Cutting wages by cutting welfare: The impact of reform on the low-wage labor market.* Washington, DC: Economic Policy Institute.

Moore, R. D., C. P. Chaulk, R. Griffith, and S. Cavalcante. 1996. Cost effectiveness of directly observed versus self-administered therapy for tuberculosis. *American Journal of Respiratory and Critical Care Medicine* 154:1013-1019.

National Center for the Early Childhood Workforce. 1994. *Training for child care jobs: Implications for low income women, children, and their communities.* Washington, DC: National Center for the Early Childhood Workforce.

———. 1996a. *Investing in child care jobs in low-income communities.* Washington, DC: National Center for the Early Childhood Workforce.

———. 1996b. Making work pay in the child care industry. Unpublished manuscript, November 25.

———. 1997a. *Making work pay in the child care industry: Promising practices for improving compensation.* Washington, DC: National Center for the Early Childhood Workforce.

———. 1997b. *NAEYC accreditation as a strategy for improving child care quality: An assessment.* Washington, DC: National Center for the Early Childhood Workforce.

National Economic Development and Law Center. 1996. *Community renewal of family economic security: The emerging role of California's family support programs in community economic development.* Oakland, CA: National Economic Development and Law Center.

North Carolina Facilities Fund. N.d. *Self help.* Brochure.

Orbovich, C. 1996. Collaborative strategies for success in the changing Medicaid market. Unpublished manuscript, April 1.

Pearl, A., and F. Riessman. 1965. *New careers for the poor: The professional in human services.* New York: Free Press.

Pickett, A. L. 1994. *Paraprofessionals in the education workforce.* Washington DC: National Education Association.

———. 1995a. An investigation of current policy, practices, and future needs in the employment and preparation of the paraeducator work force. Memorandum, October 10.

———. 1995b. Personal communication. January 19.

———. 1995c. Personal communication. February 1.

———. 1997. Personal communication. May 12.

Pitegoff, P. 1993. Child care enterprise, community development and work. *Georgetown Law Journal* 81:1897-1943.

Poland, M. L., P. T. Giblin, J. B. Waller, and I. S. Bayer. 1991. Development of a paraprofessional home visiting program for low income mothers and infants. *American Journal of Preventive Medicine* 7:204-207.

Recruiting New Teachers, Inc. 1996. *Breaking the class ceiling: Paraeducator pathways to teaching.* Belmont, MA: Recruiting New Teachers, Inc.

Riessman, F. 1984. Paraprofessionals: Twenty years later. *Social Policy* 19(3):39.

Rosenthal, E. L. 1996. *National community health advisors study: Interim report.* Tucson: Department of Family and Community Medicine, University of Arizona.

———. 1997. Personal communication. October 31.

Scallett, L., C. Brach, and E. Steel, eds. 1997. *Managed care: Challenges for children and family services.* Washington, DC: Policy Resource Center.

SEEDCO. 1996. *Community health advisors: Moving to the next level.* A feasibility study report for the Annie E. Casey foundation.

Squires, B. 1997. Personal communication. Healthy Start, Baltimore, March 17.

U.S. Department of Health and Human Services. 1994a. *Community health advisors.* Vol. 1, *Models, research and prevention.* Washington, DC: Public Health Service, Centers for Disease Control and Prevention.

———. 1994b. *Community health advisors.* Vol. 2, *Programs in the United States.* Washington, DC: Public Health Service, Centers for Disease Control and Prevention.

U.S. Department of Labor. 1994a. American workforce 1992-2005. *Bureau of Labor Statistics Bulletin* 2452(April).

———. 1994b. Occupational projections and tracking data: A statistical and research support to the 1994-95 *Occupational outlook handbook. Bureau of Labor Statistics Bulletin* 2452(May):39-44.

U.S. General Accounting Office. 1997. *States progress in implementing family preservation and support services.* Washington, DC: U.S. General Accounting Office.

Van Tosh, L., et al. 1993. *Working for a change: Employment of consumer/survivors in the design and provision of services for persons who are handicapped and mentally disabled.* Washington, DC: Substance Abuse and Mental Health Services Administration, U.S. Department of Health and Human Services.

Wasik, B. H., and R. N. Roberts. 1994. Home visitor characteristics, training and supervision. *Family Relations* 43:336-341.

Watkins, E. C., C. Harley, E. Eng, S. A. Gansless, D. Gehan, and K. Larson. 1994. Assessing the effectiveness of lay health advisors with migrant farm workers. *Family and Community Health* 16(4):72-87.

Weiss, H. B., and F. H. Jacobs. 1988. *Evaluating family support programs.* New York: Aldine de Gruyter.

Welfare recipients taking jobs often held by the working poor. 1997. *New York Times,* April 1.

Wells, N., B. Anderson, and B. Popper. 1993. *Families in program and policy: Report of a 1992 survey of family participation in state Title V programs for children with special health care needs.* Boston: Federation for Children With Special Needs.

Whitebrook, M. 1993. *National child care staffing study revised.* Oakland, CA: Child Care Employee Project.

Whitebrook, M., D. Phillips, and C. S. Howes. 1989. *Who cares? Child care teachers and the quality of child care in America.* Oakland, CA: Child Care Employee Project.

Wong, J. L. 1995. *A review of paraprofessionals in human service occupations.* Baltimore: Annie E. Casey Foundation.

Working and welfare parents compete for day care slots. 1997. *New York Times,* March 24.

Zielenbach, S. 1995. *A review of recent experience of employing local residents in social service programs.* Chicago: Woodstock Institute.

PART IV

ISSUES

10

Community Analysis and Organizing for Jobs

THOMAS R. DEWAR

Increasingly, community groups and organizations that have not previously been involved directly in "jobs" work are getting involved in it now, or are considering doing so. Others that have been involved to a limited extent are looking at ways to expand or deepen their activity. In part, this is undoubtedly due to the importance of jobs issues for many members or constituents of these groups. Even many who now have jobs have become more anxious about them, which is perhaps related to the paradoxical increase in the number of people who are both overworked and underemployed (Bluestone and Rose 1997).

For many less skilled and working-class citizens, the primacy of the jobs issue also has to do with the persistently poor performance of the current employment and training industry and of the public educational system. If people and their organizations do not feel well served by existing arrangements, they can either attempt to improve things by holding those efforts more accountable or, in some cases, get involved in actually doing some of the work themselves. Finally, some of the most effective employment and economic development programs are themselves recognizing the value of organizing and are calling for closer and more effective ties to various grassroots networks and organizations.

An increasing number of studies, national jobs initiatives, and community-building projects explicitly name better organizing as a key component (Bingham and Mier 1993; Stone 1996; Tilly 1996). By *organizing,* they refer to systematic efforts to build the voice and power of those whose condition is directly at issue—in this case, prospective workers, both unemployed and underemployed. In effect, they are saying that workers and their communities must also be at the table as these various jobs projects are designed and implemented. Too often they have not been, having been conceived of merely as clients who are dependent on whatever services their helpers are able and willing to deliver. Clients are not likely to hold programs or policies adopted "on their behalf" accountable. Organizing means these people and their communities must be considered citizens who are able to define themselves and their problems in their own terms, perceive their connections to one another, and believe in their capacity to act (Dewar 1978).

Typically, however, these calls for more and better organizing do not deal with the real perils this new effort at direct involvement in training, placement, or advocacy around jobs might hold for those on the community side. Nor do they offer the practical steps and new tools that might have to be introduced in order to integrate this kind of "jobs work" into the other important tasks community-based organizations take on. Some groups that have gotten involved describe it as "losing our way" or "getting in over our head." Others describe how they have crafted roles and strategies that seem to be working and that really add value to the mix. From these latter, more promising stories, I have selected two. Along with some themes that seem to cut across them, I will discuss the implications that community analysis and organizing for jobs of the kind represented by these stories have for project design and sustainability.

Each story represents work in progress. One is based on a strong, experienced group with a substantial track record of community building, citizen education, and public work: the BUILD organization in Baltimore. This group has managed to integrate organizing for jobs into its overall mix of strategies. The other is the story of Interfaith Action, a coalition of churches and community organizations in Minneapolis that has carried out a variety of activities in support of better jobs and community economic development that directly benefits its membership, including a Community Talent Inventory (organized as a series of conversations with individuals about their skills, experience, and dreams) and a Neighborhood Balance Sheet (through which the way capital flows into, around, and out of a particular neighborhood is studied), as well as the development of a new community-connected Hispanic Business Association.

Some framing questions ought to guide us as we consider these stories. If "community analysis" is important and consequential for "jobs organizing," just what is it and how does it develop? How do community analysis and organizing influence the development, operation, and likely impacts of the regional and comprehensive approaches being designed and put into practice?

These and the many other stories unfolding all around us have some common elements. They are all making *more effective use of residents* and their various skills, connections, and networks. They often seek to move residents from relying on the services or "good works" of others to greater reliance on themselves. They typically shift from a client to a citizen role for residents, and often from a consumer to a producer role as well. This reliance on residents is an element not just because it makes people feel good to be involved, but because it contributes to greater effectiveness. In particular, the extent and nature of social networks and personal relationships of residents are increasingly understood as key factors in their employment gains. However, it is one thing to recognize these as important and quite another to figure out how to find, strengthen, and connect them up with other activities—many of which are often more formal and more mainstream.

Aldon Morris (1984), in his study of the civil rights movement, and especially in his analysis of what set the stage for it, has noted the sustained, systematic, and largely *informal ties and social networks* created in and through black churches in the South. People seemed to be ready for action when the time and opportunity arose. This was because of work that took place "behind the scenes" or "below the radar." One of the basic questions raised by the stories presented here is how this kind of activity can be recognized, promoted, and even enhanced. Indeed, how is it possible to promote this kind of activity and informal life without distorting or eroding it in the process?

Another element found in both stories involves the value and possibilities for developing local options, strategies, and projects that *respond directly to resident aspirations* as well as to economic realities. The public work that each of these groups seeks for its citizen members to engage in also builds their "economic literacy," public voice, and citizen skills. By connecting their work to jobs and by engaging in systematic "community analysis" of one sort or another, more people become knowledgeable about what is going on around them and how they might become better connected to more promising income opportunities.

In some cases, clearly, grassroots residents also become more active and assertive through organizing for jobs; typically, they learn ways of acting together, in groups and through "public" channels, that make them better equipped to navigate the often bewildering world of work. Thus each of the stories presented here also demonstrates some of the many ways that *new voices and new roles* for citizens are being found in the often technical and expert-dominated policy debates surrounding the jobs and economic development fields.

What does it mean for people to *respect and work through community options*? Rhetoric about working more closely with the community or about developing comprehensive community collaborations that include everyone is common, but the practices necessary to bring community residents strongly into the center of the process are not. How can existing institutions begin to work with grassroots people and organizations? Further, in what ways can *existing networks, "providers," and business assets*—which already operate in the community but may not show up on anyone's official radar—be integrated into the operation of "jobs work," from identifying prospects, creating accountability, and planning power to supporting people after placement? To date, the growing interest in networks has usefully been focused on business networks (Rist and Sahay 1996), much more than on community networks. What strategies do these stories suggest might change this?

Taken together, these stories and the themes they point to can be understood as asserting a more primary role for community residents, groups, and networks. This "community first!" ethic is not an argument that other people, organizations, and connections are unimportant. It is simply an insistence that the contributions and control of residents be established. Often, this must be done before the other ingredients can be folded in. In light of what has gone before, in jobs work as in other fields, this insistence is understandable. In effect, we have long seen a style of design and operation that defines the community role as "what's left" after the professionals and their agencies have done what they can or want to do.

The new ethic, the insistence on "community first!" turns this around—or on its head. It defines the professional and institutional role as being "what's left" after the community has done what it can through its various people, relationships, groups, organizations, and networks. It is important to recognize that community analysis is not an argument against quality economic analysis; rather, in some cases, it is a necessary complement. Several factors work against more and better community analysis: Our tools for doing it are less developed, it is given less importance, and it can be at least as time-consuming as conventional analysis.

Pulling the Invisible Workforce Into the Spotlight

Baltimoreans United in Leadership Development (BUILD) began in 1978 as the response of churches in Baltimore to the pressures and concerns affecting their congregations and member families. It has grown to include 29 member churches and organizations. BUILD asserts in its founding charter, in classic Saul Alinsky style, that "power resides in either money or people" (McAfee 1993; Alinsky 1946, 1971). Then and now, BUILD does not have much money, but it does have people, thousands of them. They are all members of the churches and organizations that belong to BUILD. As an affiliate of the Industrial Areas Foundation (IAF), which Alinsky founded, BUILD reflects the major tenets of the IAF approach, including especially an emphasis on the importance of developing a program for action that grows directly out of the organizing process, a reliance on indigenous organizations and leaders, and a deep appreciation of the "self-interest" of local people.

At the very core of BUILD's vision are leadership development and community building. Through their involvement, members and leaders learn the importance of research and preparation prior to action, how to meet with power brokers who affect them, and how to create a "public life" and citizen practice through which they define issues in their own terms, determine their common interests and connections with others, and increase their capacity to act—not individually but collectively. This continuous adult education enables people to handle their own concerns and to control their own lives. "BUILD does not operate on behalf of people, but through them."

The role of BUILD staff is to support members and take direction from them. Staff are not in charge—as so often happens in resident organizations that become too dependent on their paid staff. Further, leadership is not restricted to a few, and training in support of these goals is available to all. At the center of BUILD's ongoing work are key issues that change over time but most recently have been safety and security, education, and care for the family, including children, parents, and elders.

Over the past 5 years, BUILD has also developed a Jobs Campaign. In doing so, it has also broadened its ties to other groups and institutions, most notably labor unions, especially the American Federation of State, County, and Municipal Employees (AFSCME). This reflects growing interest among both labor and community leadership for greater ties and shared work (Brecher and Costello 1990). It is important to note that BUILD did not start out to do a living wage campaign, nor did it stop its other organizing as the jobs work emerged. The organization did not do this to gain members or to impress others; it was responding directly to its members' self-interest. With so many members working in low-wage jobs, the jobs work became a natural outgrowth of BUILD's ongoing organizing, values, and practices.

Avis Ransom (1996), a member of BUILD's Strategy Team and "local leader," told me that BUILD's membership "came to jobs work gradually" and "through raw necessity." Members realized that the nature of work was changing dramatically, as was their local metropolitan economy. Although the economy had never worked very well for BUILD members, the apparent contradictions and gaps in it seemed to be growing. In addition to the widening gap between lower- and higher-skilled workers, and between where good jobs were being created and where BUILD members lived, the plight of workers without any organization or representation was clearly deteriorating rapidly. This directly affected many BUILD members (Ransom 1996).

Further, BUILD's leaders saw clearly that many of the policies and programs that were supposed to benefit the group's members simply did not. The well-documented "boom" associated with Baltimore's highly touted "urban renaissance" and Inner Harbor redevelopment had not trickled down or over to them. The jobs in these new and reinvigorated shops, hotels, offices, and restaurants around the city's new convention center and baseball stadium turned out to be poor ones, often paying only minimum wage and offering no benefits. Among the employees in these jobs were many BUILD members. Adding insult to injury, these jobs often turned out to be temporary. Coupled with an economic development strategy by the city of Baltimore that "promoted lots of low-skill, high-turnover jobs as some kind of good thing" (Ransom 1996), this trend became unacceptable; something had to be done.

BUILD and its allies took stock of this situation and organized a campaign to "pull the 'invisible workforce' into the spotlight" (Balla 1995). In spring 1993, together with AFSCME, they began to push in Baltimore for a "living wage" of $6.10 per hour for employees of contractors and subcontractors who provide services for the city. The national minimum wage was $4.25 per hour. In December 1994 they won this fight, passing a bill that raised the minimum wage for city workers, contractors, and subcontractors who did business with the city.

Along with this, AFSCME and IAF joined forces to launch a national community organizing group and created the Solidarity Sponsoring Committee/Local 1711, which fought to organize the low-wage employees of Baltimore city contractors. Today, its members make $6.60 an hour or more, and many jobs that had been "privatized" or passed over to private sector firms have been returned to government agencies.

Obviously, this is good for AFSCME, which has strengthened its base, as well as for the employee members. BUILD members believe that it is also very good for them. This kind of work, often called *workfare organizing,* keeps growing around the country, and much of it has reportedly taken heart and inspiration from the jobs work in Baltimore. "Now, Local 1711 and its 400 members have mounted a campaign to organize workfare workers, demanding that they be given all the rights of paid employees—including of course, a living wage, not a welfare check" (White 1997, 18).

According to Ransom (1996), this campaign represented "a real stretch for BUILD culture." BUILD moved well beyond its current congregational members, which have always been the focus of the group's work, and also turned to some "new tools" for organizing. As Ransom recalls, "We would show up at workplaces with coffee, juice and donuts . . . and we would just walk and talk with people." The heart of the BUILD approach is to understand "what makes people tick," to come to know "what's going on in the hearts, minds, and souls" of members. In this case, however, many of those involved were not members, and so this "listening had to be even more patient and sustained." Ransom goes on:

> We believe in the power of self-interest, and in the possibilities to develop strong leaders and a collective voice. In this case, our member churches were seeing the impact of these lousy jobs—they were literally seeing significant increases in soup lines, shelters, and church services (many of them were from BUILD member churches)—and they learned that many of these people were actually employed!

For BUILD and AFSCME, some of the issues identified, the tools used, and the relationships being built were new. This work stretched beyond both organizations' established bases. "Many of these workers were not active in churches, yet they were in our communities, in our midst" (Ransom 1996). By making the "invisible workforce" visible and heard, they also shattered the myth that these people were not capable of acting on their own behalf, or able to define their issues in their own terms, determine a course of action, and carry it out— together.

Stepping back, it is clear that although this work was new in some respects, it also grew out of a nearly two-decades-long commitment to organizing, base building, leadership development, and citizen education. BUILD had previously developed and refined the idea of a "social compact," and had successfully applied it in working with public schools, local corporations, and local officials. The organization was known and respected. Still, BUILD entered into new territory with its living wage campaign. It is very important to recognize that BUILD has a clear, strong, and farsighted vision. Even when engaging in short-term action, or fighting hard for immediate policy decisions, it is doing so with the long-term goals and aspirations of the organization in mind.

An important and intriguing new initiative that has grown out of this campaign, and one that reflects BUILD's culture and practice, is a current proposal for a cooperatively owned temporary employment service. "It will attempt to model an organization," says Ransom, who will be central to this effort, "that will pay a living wage, strongly represent the workers' interests, do quality work, and, at the same time, provide our workers with an opportunity to express their entrepreneurial sides." Through extensive community analysis and prospective client research that has set the stage for this new "temp agency," BUILD has learned that "finding and keeping good contracts" and "keeping customers happy" is very labor-intensive. Ransom adds, matter-of-factly, "Since the worker-owners are going to decide benefit levels, how to provide day care, and how to create career development and even job ladder opportunities, why not let them be directly involved in finding and keeping our contracts, too?" The vision for this new kind of temp agency is that it would be "owned, operated, and grown within BUILD culture."

Ransom explains: "We find that many of our members, and prospective members, already find themselves working in the 'temporary' sector, but very unhappily! It's time to see if we can stretch to do this kind of thing—for it is utterly clear that no one else will come in and do it for us, or do it this way—our way!" The idea is that participants will learn a variety of skills. Some may stay with the co-op, whereas others will take these skills on with them to their next opportunities, which, in most cases, will be as employees, but in a few cases might also be entrepreneurial.

Historically, much of BUILD's work has not started out with the "jobs" issue, but time and time again it has led back to jobs. Right now, BUILD has strongly integrated the issue of jobs into its overall work. "Community analysis" undertaken in support of BUILD's decisions about what to do and how to do it is not a means in itself; BUILD does not believe that facts, figures, and careful "analysis" speak for themselves. However, when they are run through BUILD's well-established culture and adult education practices, such analysis becomes a critical element in the overall mix.

To Find and Receive the Gifts of Community

Recently, the "assets approach" to community development has been popularized. An old idea being applied anew, it asserts that, to be effective, community development must start with the individual, associational, and institutional assets present in a given neighborhood (Kretzmann and McKnight 1993; Kretzmann, 1995). In this section I describe an effort in Minneapolis that is attempting to integrate this approach into its overall work.

Interfaith Action, composed of 27 churches and one community organization in Minneapolis, is a coalition that has used the assets approach as a way of getting to the issues of jobs and economic development. It, too, is an Alinsky-style organization with a strong history of success on issues of concern to residents of low-income and working-class neighborhoods in Minneapolis.

People from several member organizations, all working in the north side neighborhood of Camden, kept coming back to the issues of job loss and what was widely perceived as an eroding neighborhood economic base. Many of their members reportedly were struggling with a mix of no jobs, poor jobs, and jobs with few benefits and little future security. Leadership and staff had heard about the assets model, but were very cautious about it because it seemed too soft on organizing for power. Some of Interfaith Action's biggest successes have come through confrontation, getting power brokers to the table (or to one of their large public meetings) to account for the way the current "system" fails to work for the organization's members.

For experienced activists like these, it was not at all clear how knowing about these various assets, and then making connections among them, would benefit the members of Interfaith Action. After all, they had been trained to be wary of "outside agendas" and of "partnerships" that will not deliver for members and that they cannot control. Furthermore, this work could be very time-consuming and might distract people from their previously agreed-upon citizen action goals.

While they were trying to figure out whether and how to learn more about the various assets in their neighborhood, they commissioned an economic analysis of the neighborhood by Ken Meter of the Crossroads Resource Center. In consultation with residents, Meter (1995) developed a version of what he calls a "Neighborhood Income Statement and Balance Sheet." Relying on data from the 1990 census and a dozen other local sources, mostly official government reports, the Balance Sheet revealed much vitality in Camden, and showed significant income, business activity, and other organizational assets. It did not point decisively to what the activists should do, but it gave them plenty to think about, including some specific leverage points. It gave them a sense of "what is going on around here," and armed them with some information "of their own."

Coming at a time when many in Camden were talking of crime and job loss as the problems, the Balance Sheet contributed to a real change in focus and pointed to some leverage points that might usefully be addressed, including several small, medium, and large employers that had little or no connection to local residents or to local suppliers. It also showed a drain of resources out of the neighborhood, via the millions of dollars of mortgage debt retired each year through the much higher rate of taxes paid than of services/benefits received by those same taxing bodies. It revealed an array of associations no one seemed fully aware of before, an employment base larger and more diverse than previously thought, and estimated barter activity large enough to warrant serious discussion.

But here as elsewhere, information alone is not enough. What is the likelihood that these various assets can be tapped, and how might they be connected in some concerted way? Does information available through these "official channels" tell you what you really need to know, even if it has been packaged for neighborhood use by someone like Meter? Probably not. The people working in Camden decided they wanted more dialogue with neighbors, strangers, and friends.

The assets approach, as promoted by Kretzmann and McKnight, is based on the idea that we typically do not know as much as we ought to about the individual gifts and talents of the residents in our communities. Too often, we know nothing at all about them, have preconceptions or prejudices about them, or look at them in terms of their problems or needs. In the process, we completely miss the myriad possibilities represented by their gifts. Further, we also have many community associations all around us, from clubs and choirs to youth, senior citizen, and neighborhood groups. These rely primarily on the unpaid work of members and are self-directing and self-sustaining. Rarely, if ever, do we find out much about groups to which we do not personally belong. What do they really do? Who is in them? Do they have a sense of "public work"? Are they open to being involved with other groups in common projects? These are the kinds of reasons activists should visit and learn about both individuals and community associations.

Still, for groups accustomed to "organizing for power," this sort of visiting with "strangers" can seem soft or vague. And sometimes it is. Overcoming their cautions, the members of Interfaith Action developed and launched the Community Talent Inventory, which focused on residents of the Camden neighborhood. Four member organizations within Interfaith Action—three congregations and one community organization—identified a team of members to help design and carry out the process, first to learn about how to do these visits and then to train others, and to assist in deciding what to do with all the information once it was collected. Each group operated slightly differently from the others. Some focused internally on their members. Others tried to visit with neighbors who lived close to their churches while also interviewing their members. Still others, intrigued by certain informal networks active in the community, tried to check in with people active in those.

A common form was developed and was then adapted by each group through the addition or deletion of questions. A total of 60 interviewers were eventually trained and sent out. Possible uses for the inventory results were discussed as the work got under way, and these conversations continued as people came in with their newfound insights about "who lives here." This ongoing discussion of purposes was later seen as very important—it motivated those going out, put some practical ideas in their minds, and allowed them to answer more effectively the inevitable questions of the people being interviewed, such as, What are you going to do with this? What is going to happen next? In this case, economic development, community-based (and home-based) business development, and job connections were the focus. Knowing this affected how the interviewers listened, by making them active, and influenced what they followed up on most. The interviewers were prepared with probes to use as well as information on how home-based businesses or barter networks evolve.

Over a 4-month period, information was gathered from 260 Camden-area residents in a series of one-to-one conversations, in small groups at homes and in churches, and through a series of Sunday morning "pew interviews" in which time was set aside for the congregation to interact around these issues. The methods varied according to the experience and best judgment of the sponsoring groups. What would work here? How can we get people to talk openly about their dreams, aspirations, skills, and job-related experiences?

There were multiple audiences for this "community analysis." Importantly, each group sought to integrate this activity into other ongoing work, dialogue, and relationship building. The information gathered did not speak for itself. It was not a stand-alone strategy. Different groups understood and used the talent inventory in different ways. In one area, information from 100 interviews showed that a remarkable number of people wanted to become more computer literate and skilled. This has led to the creation of a computer literacy center for skills development and skills exchange. Local residents with the right kinds of experience and willingness to share what they know, some space, and used but still recent computers were gathered to launch this center.

In another area, driven by the common interests of Spanish-speaking immigrants and spearheaded by an emerging new congregation, Sagrado Corazón, that is working through them, the results of 60 inventories were used to guide job and enterprise development. A number of Hispanic-owned businesses already exist in or near the area, but they are not well publicized and were not previously working with each other on common issues. Some of these issues, such as difficulty in getting financing and culturally appropriate technical assistance, created natural bridges between the new group and the core strategies and base of Interfaith Action's traditional organizing model. A development team has formed to develop a business incubator—*mercado*— and a festival that is planned for the summer of 1998 to celebrate and deepen ties to the talents and skills of the Hispanic community in Minneapolis. This all began in the Community Talent Inventory, in one neighborhood, but has quickly broadened out in search of the kinds of metropolitan (or regional) connections that viable economic sectors and networks create and maintain (Clark and Dawson 1995; Melendez 1996). It has also led to an expansion of the inventory into four other member congregations, each of which is planning its own survey as of the time of this writing.

In the process of doing this "community analysis," these groups have learned a great deal about themselves, and new connections among members have been made. But perhaps just as important, they have become much more knowledgeable about the existing "employment and training" system. As one leader told me, they have been led to ask much more forcefully, "Why can't such talented people as we have here find better jobs, and more income? Why can't they get small business loans in Camden? Why isn't more appropriate technical assistance available?" The group organizing the *mercado* has also proceeded to do work around the poor performance to date of several city and state programs claiming to promote "minority entrepreneurship" and small business development, but with which community members have had almost uniformly bad experiences.

In effect, the Camden community has been invigorated by the kind of "community analysis" that led to and has resulted from the Community Talent Inventory. This has made community members more inclined both to do some selected things themselves and to engage in more pointed and strategic planning about holding existing providers accountable for who they serve and how well they serve them. This is a useful supplement to their traditional organizing practices. Occasionally, they do find institutions and programs with which to work, but these relationships are results oriented. They are held accountable for what they claim. Sal Miranda, lead organizer for Interfaith Action, says that these efforts are more like "negotiated working agreements" than "public-private partnerships."

Although leaders were open to finding surprises and new leads, the clear emphasis from the outset was on creating more jobs, supporting and growing local small businesses, and keeping dollars recirculating within the local community economy. They found many skills, talents, and even business interests, but the real challenge turned out to be less in finding the gifts and more in how to put them to better use. The experience of Interfaith Action, as well as the experience of many other groups, suggests that receiving the gifts remains the most difficult and largely uncharted part of the "assets approach."

Re-Membering Community

The stories presented above remind us of the key role that strong organizing for jobs and appropriate community analysis can provide. They are too often thought of as separate from or even in opposition to the design and implementation of effective employment strategies. When trying to stand alone, neighborhood strategies cut themselves off from the kinds of connections and better jobs they need. Geography alone can be a limiting factor, but geography remains one of the most common elements in how people join and form groups. It is these *permanent membership groups,* in turn, that give any community its capacity to act in self-directed and independent ways—whether that community is defined in terms of place, shared condition, common workplace, or something else.

These three stories tell us about more than just the themes identified at the outset of this chapter. They also reveal the extent to which groups are getting involved in jobs work and community analysis that adds real value to the mix, strengthening the overall results and democratizing the process. Their critique, usually very clear, is as important as their vision.

For example, the Reverend Johnny Ray Youngblood (1996), a leader and strong community voice with East Brooklyn Congregations in East New York, has described three approaches he believes just "don't work." Both of the organizations represented by these stories share a skepticism about the many conventional approaches often recommended to them. One of these is the "outside-in response," which says "that cities and regions can only be saved by those who don't live within them—by tourists, shoppers, sports fans, theater-goers, casino gamblers, and conventioneers." This response calls for cuts in social spending and hikes in subsidies and tax breaks to build the business and entertainment centers capable of attracting those who live (and spend) outside the cities. People who promote this response see cities as museum districts, profit centers, pleasure places—"their public faces grinning at commuters, their back to their own citizens." Although more challenged than it once was, this response is still common.

A second response that doesn't work is "more and better programs." To focus too narrowly on what can be done solely within programs alone (however much improved) is to misunderstand their well-documented ineffectiveness to date. The fact that many have received their resources in the name of their clients—the intended beneficiaries—and have "run them down" as incompetent, ignorant, or impaired in the process is only that much more offensive.

A third response that Youngblood says does not work is "more public-private partnerships." These usually involve the "private sector and the government sector," with a token community advocate or preacher on the board for window dressing. This kind of response tends to be tame and nonconfrontational. "Let's not point fingers, we're told; let's get something done. It tends to be led by professionals and planners, to be acceptable to funders and grant makers, and not to have very much impact."

So then, what's the response, for Youngblood and for others? What do these stories suggest groups and coalitions are doing? First, start inside, not outside. Start with who and what are close at hand: the people, associations, and local institutions that have forged good community connections. Second, look for ways to make work pay. Third, increase the opportunities for ownership and equity. Identify and increase the long-term stakes for grassroots residents in how things work. When they find and use their voices, these are common messages.

"Community analysis" is not a strategy in itself, but these stories suggest it is often an early step in putting grassroots residents at the table. It also points to what they bring, as well as to what they must strengthen or find elsewhere. And organizing is not the same as recruiting. Much too often, grassroots residents are tolerated or patronized rather than welcomed—it's politically correct. Their support is sought, but not their ideas, experience, connections, or energy. When this is the case, they are being recruited to help carry out the ideas of others, rather than to develop and implement their own.

Good organizing builds the membership base, brings people into groups, and brings groups into networks. It is always looking for ways to increase impact and improve results—as measured by the benefits residents actually experience. This explains much of the growing interest in jobs work among a variety of grassroots groups.

Unfortunately, jobs work and, more broadly, community economic development in general are full of "camps." Each is prone to use its own language and logic to describe what it does and to report its wisdom or insight to others. Too often, people operating in these various camps talk past one another, and rarely, if at all, do they have dialogue with the people who are supposed to benefit from the policies, projects, and programs with which they are preoccupied. Grassroots organizing is also one of these camps.

Social networks, so often noted as a key ingredient in job finding and keeping, are not so easy to build or to pull over to someone else's purposes. Strong networks are natural phenomena. They can be created, or strengthened, through self-conscious support, but at their heart they reflect what the participants in them see as their interests. They are largely self-governing, and seem to resist being enlisted for some "wholly different cause or purpose" from what participants see as the reason to be connected. Most of us know that our social relationships, our connections, are the keys to our finding and keeping good work. Indeed, this has become a more and more explicit element of strong jobs work (Melendez 1996), whether that work is programmatic (as in the case of the Center for Employment Training) or community based (Stone 1996). If others are trying to tap into existing networks with the goal of having them add jobs work to their agendas, there will be considerable caution by those involved about sharing or opening them up to others (Taub 1990). The very idea of enlisting social networks and putting them to work on broader community issues is offensive to many in those networks, and reveals how little we know about how social networks form, operate, evolve, and fade away.

These stories also represent a mix of strategies: from a single "neighborhood" (composed of multiple and overlapping "communities") whose members are trying to focus in more deeply where they are, connecting to broader jobs connections and possibilities; to citywide policy advocacy and enterprise development based on a solid base of citizen members and track record; to a regional vision, drawing on deeply felt values and aspects of local history and culture, that includes many who are traditionally hard to "organize" (e.g., craftspeople, and the social and educational infrastructure that supports and works with them). These are only a sample of what is to come, for the inclination and the capacity for grassroots residents to become more directly involved in jobs work is only going to increase.

Re-membering community requires greater direct involvement on the part of grassroots residents. Just how they might best be involved in jobs work is less clear, although patterns are becoming clearer. The emphasis they bring on finding systematic ways for themselves to contribute, belong, cooperate, and control what happens are all healthy. Serious discussion about both the problems and prospects for this work, based on the kinds of impacts, results, and limitations it demonstrates, would be helped by more and better accounts of what is being tried in various communities.

References

Alinsky, S. D. 1946. *Reveille for radicals.* Chicago: University of Chicago Press.

——. 1971. *Rules for radicals: A practical primer for realistic radicals.* New York: Vintage.

Balla, L. 1995. Baltimore's wage war. *Neighborhood Works,* February-March.

Bingham, R. D., and R. Mier, eds. 1993. *Theories of local economic development: Perspectives from across the disciplines.* Newbury Park, CA: Sage.

Bluestone, B., and S. Rose. 1997. Overworked and underemployed: Unraveling an economic enigma. *American Prospect* 31 (March-April):58-69.

Brecher, J., and T. Costello. 1990. *Building bridges: The emerging grassroots coalition of labor and community.* New York: Monthly Review Press.

Clark, P., and S. L. Dawson, with A. J. Kays, F. Molina, and R. Surpin. 1995. *Jobs and the urban poor: Privately initiated sectoral strategies.* Washington, DC: Aspen Institute.

Dewar, T. 1978. The professionalization of the client. *Social Policy* (January-February):4-9.

Kretzmann, J. 1995. Building communities from the inside out. *Shelterforce* (September-October) 8-10.

Kretzmann, J., and J. McKnight. 1993. *Building communities from the inside out.* Chicago: ACTA.

McAfee, N. 1993. Relationship and power: An interview with Ernesto Cortes, Jr. *Kettering Review* (Summer).

Melendez, E. 1996. *Working on jobs: The Center for Employment Training.* Boston: Mauricio Gastón Institute, University of Massachusetts.

Meter, K. 1995. *Camden neighborhood income statement and balance sheet.* Minneapolis: Crossroads Resource Center.

Morris, A. D. 1984. *The origins of the civil rights movement: Black communities organizing for change.* New York: Free Press.

Ransom, A. 1996. Interview. R&B Limited, Baltimore, December 16.

Rist, C., and P. Sahay. 1996. *Community-based organizations and business networks: New ideas for creating job opportunities for inner-city residents.* Washington, DC: Corporation for Enterprise Development.

Stone, R., ed. 1996. *Core issues in comprehensive community-building initiatives.* Chicago: Chapin Hall Center for Children, University of Chicago.

Taub, R. 1990. *Nuance and meaning in community development: Finding community and development.* New York: Community Development Research Center, New School for Social Research.

Tilly, C. 1996. *The good, the bad, and the ugly: Good and bad jobs in the United States at the millennium.* New York: Russell Sage Foundation.

White, A. 1997. WEP: Workers expect paychecks. *City Limits* 22(3):13-19.

Youngblood, J. R. 1996. A call for organizing, confrontation, and community building. *NFG Reports: The Newsletter of the Neighborhood Funders Group* 3(1):3-4, 9.

11

The Politics of Jobs in Maine

Pierre Clavel
Karen Westmont

That jobs, and more particularly good-quality jobs, have gotten onto the agenda of state and local political discussion at all is worth a lot of notice. For the most part, political sentiment still insists on business prerogatives for the supply of work. The role of the public sector is to be as helpful as possible. Private interests have usually opposed such interventions as the minimum wage or worker buyouts of failing mills. In the face of such opposition, public servants almost always retreated, making the timid calculation that, after all, top management drives the economy and jobs. That calculation was not easily questioned, and, above all, no one wanted to scare jobs away.

This reluctance to bargain with the private sector continues, despite an increasing list of arguments to the contrary. C. B. Macpherson (1977) argued on political theory grounds that democracy itself was at risk unless increased citizen participation created a setting for debate with private interests. One likely avenue—the self-interested participation of disadvantaged and lower-income segments—was blocked in various ways. One of the most important blockages was the traditional failure of labor and community groups to reinforce one another, or even to make much contact (Katznelson 1981). The remedy seems to be to encourage such contact, but so far little progress has been made on this or other approaches either to greater participation or to more mobilized public opinion.[1] Several local efforts suggest that the historic bifurcation of labor and community is changing, and that support for public intervention to increase the supply and quality of jobs has begun to appear. One of the best known was the 1995 campaign for a Living Wage in Baltimore (see Dewar, Chapter 10, this volume). A succession of similar campaigns or actions in at least 24 states of the nation has so changed the political debate and public opinion that they deserve to be termed a "Good Jobs Movement."[2] Win or lose, these campaigns for quality jobs illustrate the potentional for a new community standard of who reaps the benefits of work and of public investment in jobs.

The Politics of Jobs in Maine

The good jobs campaign in Maine both mirrored the general flow of events and was in some ways different—behind some places, ahead of others, and in some respects unique.[3] By looking at one state in some detail, however, we may be able to understand better what is happening in the nation more generally.

Maine in 1997 was a generally rural state facing the challenge of job loss and the transition from a branch plant and natural resources economy to one that would increasingly depend on small manufacturing start-ups, tourism, and retirement services. It had a small population, a homogeneous ethnic makeup, and both a tradition of independence and a history of dependence on large and absentee employers. But it also had—in common with other New England states—vibrant and innovative demographics, as new populations moved in and worked creatively with the long-term residents.

The issue of job loss and creation had been prominent in Maine for a long time, and by the mid-1990s was joined by concerns about job quality, focused by the programmatic work of a number of groups. Historically, the state government focused on the attraction of large branch plant manufacturing, and there was an economic development culture that supported that function. The state planning office tried to predict the sectors most probable as targets for attraction strategies, organized labor tried to defend relatively higher-paying manufacturing jobs against layoffs and plant closings, and the public finance and subsidy strategy avoided aggressive efforts to intrude on private labor relations issues or to hold companies accountable for performance levels on job retention or job quality.

By 1997, however, this historical picture was changing. Plant closings had undercut the attraction of the branch plant model of economic development, persistent rural poverty had stimulated a set of innovative experiments, a widely publicized case of worker mistreatment at a poultry farm near Lewiston captured public and press attention, and other events put state economic policy on the political agenda in the legislature.[4] There was a vigorous campaign to impose accountability on job providers and to pass a minimum wage law.

Official Economic Policy

Maine had long been able to build up its personal income and profits by attracting manufacturing operations. Its chief instruments were the Finance Authority of Maine (FAME), which had financed hundreds of manufacturing and other enterprises, and the state planning office, whose research and site inventory operations backed up FAME's efforts.

The 1994 gubernatorial election of independent Angus King brought new life to the economic development operation. King met often with representatives of business; he "touted the three Ts—technology, trade and tourism" and unveiled an economic development strategy to shift the focus away from branch plant manufacturing. Only gradually did the new governor begin to fill in the details, however; obviously, he was trying to respond to contending visions of what might be appropriate. Apparently the state was not going to stop subsidizing major branch plants.

Maine's State Department of Human Services (DHS) policy complemented the branch plant attraction approach by limiting its work mainly to servicing welfare clients and avoiding extensive support for workforce training initiatives. When the federal welfare reform bill passed in 1996, the state DHS commissioner focused attention on putting welfare clients to work, regardless of job quality. The argument from DHS was, "Any job is better than no job."[5]

Some of the good jobs ideas did get into official channels in the early 1990s. The state changed a supplementary tax increment financing instrument to require wages equal to those prevailing in the area as a condition for the subsidy. Most important, the Maine Economic Growth Council, which was cochaired by State Senator Rochelle Pingree and Kevin Gildart of Bath Iron Works, generated a series of subcommittee discussions in 1995 that put the good jobs standard into play and resulted in its appearance in two of six subcommittee reports (Maine Development Foundation 1996a, 1996b). The Growth Council report was particularly important, as it provided for annual reporting of progress toward quantifiable goals, including the numbers of good jobs in Maine firms.

The Critique of Official Policy

Most striking was the way official policy came under attack and an alternative program emerged during 1995 and 1996. David Vail and Michael Hillard, economists at Bowdoin and the University of Southern Maine, who later would argue at length for the provision of good jobs, began with an editorial in the newsletter of the Maine Center for Economic Policy (MCEP), an economic advocacy voice and think tank supported by national foundations (Vail and Hillard 1995). They noted the dangers of the governor's initial statements at the beginning of his term in January 1995: "As the Governor prepares to reveal details of the strategy, citizens should think critically about an economic vision with business needs and profits—rather than people's needs and well-being—at its center." They argued for a focus on high-quality jobs; for adult education, training, labor market services, and regional targeting of high-unemployment counties; and for a focus on those least well-off.

By 1996, attention to the plight of Maine's poor and working-class population had made its way into the state's mainstream press. When National Semiconductor, after receiving subsidies from FAME, announced job cutbacks, a *Morning Sentinel* op-ed piece titled "King Has Been Bamboozled. National Semiconductor Reaps Tax Break, Job Cuts" (Williamson, 1996) inquestioned the policy of unconditional tax breaks. Other writers questioned the tax increment financing policy that gave subsidies to businesses with no guaranteed benefits (Halper 1996), and a *Bangor Daily News* editorial echoed Vail and Hillard in a call for the state to target the high-unemployment counties ("Bull's-Eye of Success," 1996).

The Maine Economic Policy Center kept up the pressure: "Christopher St. John said he generally supports King's plan as a starting point. . . . Still, he is concerned that, if carried out, it will boost business profits but will not create enough high-paying jobs or help struggling regions within the state" (Porter 1996).

Press attention began to emphasize the plight of Maine's poor, and focused on the state's initiatives (or lack thereof). The most notable of many accounts was a seven-part newspaper series by Eric Blom and Andrew Garber. They reported increasing poverty and inequality, and went on to cite policy ideas and trial balloons that were making their way onto the public agenda in 1996. First, the governor seemed to be picking up some of the rhetoric of the critics: "Governor Angus King has suggested tying tax breaks for some businesses to the hourly wage paid by a company. The higher the wage, the bigger the tax break. 'I try to concentrate on the better jobs,' King said during an interview for this series. 'We have limited resources so what I'm concentrating on is manufacturing jobs and value-added jobs' " (quoted in Blom and Garber 1996).

Others, like Senator Pingree, went further, suggesting a "corporate accountability" theme: "In everything else we do in government, everyone expects us to have standards to be measured by," she said. "Business's performance at producing good jobs also should be measured." The Blom and Garber series also described State Senator Sean Faircloth's proposal of a five-point plan to address the growing income gap, one provision of which would label companies that meet certain criteria as "responsible corporate citizens." These companies would pay an 11% corporate income tax rate, compared with an 18% rate for other companies. There were also proposals from democratic legislators to create health insurance alliances for small businesses and to make insurance portable.

Republicans, who had gained the legislative majority with the national sweep of 1994, were by now in a quandary. Blom and Garber (1996) reported that "Republicans tend to want to avoid any singling out of firms, and so oppose these targetting ideas. John Hathaway commented 'Any job is a good job that is created where there wasn't one before.' Susan Collins supports broad based lower taxes; Olympia Snowe advocates budget balance."

Organizing

The press attention was part of a national phenomenon. The Blom and Garber series, for example, may have been modeled on an earlier one by *Philadelphia Inquirer* reporters Barlett and Steele (1992), and was part of the shift in public opinion as the national Republican "revolution" wound down. There was also, however, some astute organizing in Maine's liberal community. The Maine Center for Economic Policy was created in 1994, when the Ford and Mott Foundations, perceiving that state fiscal crises had tended to be resolved in favor of the affluent, sought to stimulate citizen participation; if citizen education could be improved, they reasoned, then legislatures might take a more enlightened role. The organizers of MCEP took as a model the national Center for Budget and Policy Priorities; however, they took on issues of infrastructure and jobs as well as fiscal issues (St. John 1996).

Social services agencies, meanwhile, saw the challenge they would face after the Republican victories of 1994. Chris Hastedt (1996), director of the Maine Equal Justice Center and a key organizer and lobbyist, related their strategy:

> When the welfare reform mania hit a couple of years ago, we started a campaign to shift the debate. Our modest goal was to get people to question the stereotypes

about welfare mothers. We did a survey of 3,000 AFDC families, got a 33% return; we hired Stephanie Seguino [1995] to do a report, "Women Living on the Edge." We did press conferences. . . . We were able to defeat most of the bad stuff.

We needed to do some positive stuff. Sharon Treat proposed a bill in the legislature. All we got was the Commission on Hunger and the Commission to Study Poverty Among Working Parents. This did produce an agenda.

The *Report of the Commission to Study Poverty Among Working Parents* did make positive recommendations at the end of 1996, on topics such as tax credits, child care, and minimum income standards for low-income workers (Maine State Legislature 1996). Hastedt next hoped to "change the climate about [employment]," meaning how jobs were structured in the workplace. This would be another kind of policy initiative, supplementing public measures with pressure on private employers.

Legislative Initiatives in 1997

By the end of 1996, the liberal comeback had gained momentum. The national welfare reform bill had passed the initiative to the states, and passage of the federal minimum wage had been a modest victory for liberal forces. In Maine, a series of measures was on the agenda. The Democrats regained the majority in both houses of the legislature. "Corporate responsibility" and "good jobs" were on the agenda. Rochelle Pingree, Senate majority leader, was one of several point persons for liberal groups. Her assessment was that the climate had changed, and new legislation was possible:

> When I came in the legislature 4 years ago, government was the only pond scum around. The attitude on economic development was (a) business was good, the main engine of development; and (b) don't do any "social contract"-type stuff tied to subsidies because it might scare them away.
>
> That atmosphere has changed now. There have been these series in the press on the inequalities between "rich and poor." When I go back to my district, I meet with a lot of small businesses, because that is mainly what there is. But I can't talk to them about something like "We just gave a subsidy to Champion Paper." They say, "Yes, but what about us?" They see a big difference now, they discriminate. (Pingree 1996)

On Pingree's desk were several proposals that came from the activism and studies of the previous 2 years. Some form of legislation to contract "corporate accountability" was at the top of the list. Another concerned skills training for low-income women and reorganization of the job training system. Kit St. John (1996) thought there would be at least a good debate in the legislature around a proposal to raise the state minimum wage over the federal level.

Only a few of these initiatives would bear immediate fruit. The Department of Labor did a superficial reorganization of part of its training programs, called the "Governor's Training Initiative," without affecting the far larger potential financed by federal monies. A vote on the minimum wage, after committee discussion, was held over until 1998, when the governor—who opposed the increase—would be up for reelection and less likely to veto it. "Corporate accountability" proved complex; a bill was withdrawn with the explanation that the issue demanded further study.

The resistance to "good jobs" legislation would be substantial, and a cautious approach in the legislature seemed justified. Though Democrats had the majority in both houses, and considerable sentiment for reform, the governor straddled the fence between protection of business interests and popular demands for accountability and "good jobs," and remained opposed to many of the Democrats' ideas. King had come far enough to acknowledge that inequalities between rich and poor workers, and between growing and lagging areas, were a serious problem, but not far enough to advocate any specific policies to address them.

Alan Brigham (1997), a gubernatorial appointee to the Department of Economic and Community Development, expressed the resistance to such policies: "I'm a business developer. We do want high-road employers. But we also realize that the high road is not for everyone. High-road practices should not be a condition for state support. We should respect that some firms will have other management philosophies." Workplace audits to examine "good jobs" and employee relations would be valuable, he thought, but not if they were used to impose any standards on management: "We tread on dangerous ground when we start tying our support to specific management principles. Businesses need to manage as they see fit."

New Capacities Emerge

The tension between a strong legislative agenda for "good jobs" and a governor and state agencies that sought to maintain what they thought was useful in past practices and relationships that favored business and agency prerogatives continued into 1997. However, the "good jobs" focus promised to continue into 1998, through the legislative and gubernatorial term. One factor was the pressure that would come from continued efforts to put welfare recipients to work.

Another factor that seemed to ensure continued pressure for good jobs was the experience of several concrete initiatives that had been going on for some time—as long as two decades—and that were showing increased capacity to deal realistically with economic problems. First, there was the movement to focus attention on the value of high-quality workplace organization. Second, there was a continued stubborn resistance to labor losses, including a renewed capability to work with business, government, and nonprofit groups in response to corporate plant closings. Third, there was a series of women's workplace initiatives. Finally, a series of innovative training initiatives tied to finance instruments promised to have increased relevance as the welfare-to-work pressures increased through the end of the decade.

■ High-Quality Workplace Movement

The idea that "good jobs" rather than just "jobs" ought to be a matter of state policy was emerging out of the practice of at least some businesses as well as deliberations of the Economic Growth Council. In May 1997, economists Hillard and Vail convened the "Taking the High Road . . . " conference at the University of Maine in Augusta to air the thought that state economic policy ought to take the "high road," favoring firms that were "high-performance workplace organizations" (HPWOs) over low-road firms.

They noted, as many others had, the increasing inequalities between the rich and poor, and between the industrial southern counties and the more rural parts of Maine. They then argued that businesses, not government alone, needed to right these problems. The key, they thought, was to promote HPWOs. They had written a paper describing some of these firms. The CEO of one of the firms, Ducktrap River Fish Farm, provided a list of principles that begins to define the HPWO:

> A cared-for employee is more productive than one that is less cared for.
> Every employee has untapped potential.
> We spend a majority of the day at work, so it has to be a place where people can grow.
> A company that can't tolerate mistakes will not be as effective. The larger culture is into shame and blame. Business should counter this.
> You have to recognize good work, through profit sharing or evaluation procedures that confer ownership. (Fitzgerald 1997)

This agenda was followed by a presentation from Champion Paper, a large mill in Bucksport that had instituted a participatory management initiative in the mid-1980s. Despite moderate downsizing, there was little labor-management conflict at the company; the keys were the participation of union membership, an agreement that no jobs were to be lost through layoffs, and a commitment by management to invest in training.

The conference presenters noted that there were dozens of these sorts of high-performance workplaces in Maine. Though such businesses represented only a small portion of Maine firms (5% seemed a good guess, though no one had done a census), the organizers argued that public support for these businesses was the best way to attack problems of poverty and regional disparities. They presented this "demand-side focus" in labor market policy (invest in HPWOs) as the necessary complement to a "supply-side focus" (invest in people). To pursue such a policy, they suggested four steps:

- *Build on success.* Create a consortium of HPWOs. Start with audits and job quality awards. Award-winning firms would be models and mentors for other firms. This is as much a matter of inspiration and encouragement as technical assistance. Supplement with short courses. Focus on lagging, low-road sectors. Extend the focus to microenterprises of fewer than 25 employees.
- *Encourage rural businesses to adopt HPWO practices.*
- *Build workplace reform into all state-supported business development programs*—and get them out into the countryside.
- *Hold businesses accountable* for job quality if they receive public assistance.

Hillard and Vail knew that what they were proposing would be controversial. Their third and fourth proposals had received negative comments in op-ed pages, and prompted the remarks by state official Alan Brigham quoted above. But they also were specific enough to connect with other groups and with more general policy positions. The idea of workplace audits received general approval—there was already an organization composed of adherents of HPWOs within business and academia, the Maine Quality Center, which provided audits on a largely voluntary and free basis, and was projecting 80-100 applications for audits annually by the year 2000.[6] The idea of corporate accountability had a relatively large, if unspecific, public presence. In fact, the workplace audit idea, by specifying what corporations should be accountable for, may have triggered the vehement defensive response to Hillard and Vail's report.

Vail, summing up the discussion, emphasized the positive response to the idea of audits and the possibility that fervor for quality might push the HPWO initiative from below, as much as policy pressure from above would enforce it (one op-ed writer had caustically referred to the "quality police"). And it would be hard to exaggerate the intensity with which HPWO enthusiasts advocated their cause and sought to transfer their principles to other firms. In one question-and-answer period, an audience member suggested, "Don't call it *quality management*; a better term is *quality fever.*"

■ Labor Buyout Capacity

Labor had been strong in paper factories and in industrial areas until the 1980s, but a series of setbacks, particularly a lost strike at the International Paper Company in Jay in the late 1980s, had weakened it. Its membership had slipped as a share of the labor force by 1996. But labor shared in the liberal comeback after 1994. It had some modest legislative victories, such as a 1997 law mandating overtime pay for industrial poultry workers, and it was pressing for more generous treatment on workers' compensation and a state minimum wage. Most notably, the state's labor movement had developed the capacity to participate in employee buyout deals as a response to plant closing pressures (Cavanaugh 1997).

Maine manufacturing had a modest history of worker buyouts. John Roberts, a small clothing factory in Biddeford, was saved temporarily by an employee buyout, but went bankrupt after a short period. In 1996, the well-known C. F. Hathaway shirt factory in Waterville, which had been acquired by the conglomerate WARNACO, was up for sale and threatened with closure. Mike Cavanaugh, northern New England director of the Union of Needletrades, Industrial and Textile Employees, who had been part of the Roberts effort earlier, helped put together a group of investors that ultimately purchased the plant, saving 450 jobs. The employee buyout, which had been part of the plan at one point, never materialized, as other investors, including Alice Walton, heir to the Wal-Mart fortune, and former governor William McKernan, stepped forward.

In November 1996, Sunbeam Corporation announced that it would sell its subsidiary Biddeford Textile Company, which made the woolen blanket shells for Sunbeam's electric blankets. Cavanaugh, fresh from the Hathaway buyout experience, went to work again. He later said, "We took the position that rather than wait and figure out who's going to come in and buy the plant, we needed to take the initiative. After all, it was our jobs on the line" (quoted in Beaudoin 1997).

He worked with Rene Boisvert, who had been the Biddeford CEO, and contacted Michael Liberty, a Portland developer who had been instrumental in putting together the Hathaway deal and who again involved Alice Walton. This time Cavanaugh was successful in creating a worker buyout. He prevailed upon the investor group to bring in American Capital Strategies, a firm that specialized in worker buyouts, which did a feasibility study that persuaded the union membership to agree to the move. The local union in the plant became a one-third partner in the buyout, which went ahead with a 5-year guaranteed contract for its output from Sunbeam, its former parent corporation.

■ Women's Economic Initiatives

In Maine as elsewhere, one of the big stories was the growth of civic life through the emergence of women's organizations. Women took larger roles in the labor movement; there was the work of Women Unlimited in training workers and advocating for apprenticeships in the trades, and the Women's Business Development Corporation, which helped further the entrance of women into business as entrepreneurs. Earlier than these was the growth of Maine Displaced Homemakers (MDH), later Maine Centers for Women, Work and Community (MCWWC) (Nardone 1997). Their key organizers had been involved in advocacy and networking nationally starting in the 1970s, and in 1977 the legislature in Maine appropriated $15,000 to establish a position at the University of Maine at Augusta that would assist women making the transition from home to work. MDH found fertile ground working with women who were experiencing separation or divorce and who needed to develop skills and an approach to the workplace. It offered one-on-one counseling as well as workshops and training classes on such topics as personal empowerment, entrepreneurship, workplace skills, and community leadership. By the 1990s, MDH was operating from 13 centers throughout the state and working with about a thousand women each year.

During the 1990s MDH diversified. It did more targeted training, providing applicants for placement in specific factory jobs and off the welfare rolls. Seeing that the jobs its enrollees had access to were too temporary and dead-end to support families, the organization became involved in economic policy advocacy, making it part of the developing "good jobs" movement in Maine and tying it more closely to other organizations working on similar issues. Some of this happened informally, as when MDH organizers made common cause with women involved in the mid-1980s strike in Jay, and later formally, as when a foundation board member encouraged the organization to set up a course for the women workers at Hathaway Shirts, and MDH arranged for union organizers to address the group.

- Employment and
  Training Agreements

David Vail (1997), in summing up the "High Road" conference ideas, carefully mentioned the more subtle approach to "public-private partnerships" that would be necessary to create high-performance work organizations. They would have to include not only government and business partners, but also labor and community-based organizations. He might have gone on to elaborate: If changes were to occur in the workplace, it would be necessary to have a unit in place specifically dedicated to making those changes happen. One of the most effective examples in Maine was the employment and training agreement (ETAG) approach to business subsidies that had been developed by Coastal Enterprises, Inc. (CEI), a community development corporation in Wiscasset (Clavel 1996). The ETAGs emerged in the 1980s when that organization decided to use its already developed grant-raising abilities in the interest of small manufacturing businesses. CEI's director, Ron Phillips, saw that it would be possible to use training funds as subsidies, and tie them to business loans, if CEI could maintain a role monitoring the relationship between the firm and the training agencies (normally directly under the supervision of state agencies). CEI was, in effect, delegated for what had been a supervisory role played by the state. It was a form of decentralization through contract.

Phillips reasoned that small businesses undergoing expansion needed to minimize the effects of capital costs and maintain cash flow, but development finance specialists had overlooked employment and training programs

> as a source of cash relief.... Wage subsidies for trainees and tax credits for hiring the disadvantaged are just two resources which can free up an expanding business's cash whether to cover increased debt service on new buildings or to purchase critical additional inventory. In the long run, employment training programs pay off in increased productivity and enhanced product quality, but their immediate impact on cash flow can be as positive as a low interest loan and just as important to insuring that a business survives the cash "crunch" caused by expansion. (Clavel 1996, 28)

CEI proposed that it be the key link connecting the development finance function (business loans) and the training function, previously carried out by state Department of Labor contractors, but not effectively tied to targeted low-income populations such as AFDC recipients, or to as many private firms as might potentially become involved in training efforts. CEI proposed a three-way relationship involving the training agency and contractors—and the state Department of Human Services, where welfare clients were to be involved—the private firm seeking expansion funds, and CEI as monitor of the relationship.[7]

The key CEI role was monitoring. CEI was convinced that the training programs would never work without CEI's taking this role:

> Our role is that we are there monitoring. If we weren't, if we just did the agreement and then walked away, believe me, nothing would happen. If we see a classified ad we hadn't been notified of beforehand we are on the phone immediately, reminding them that the agreement says we are to be notified before any ads appear. There is a constant push on the system, more on the state agencies than on the firms, which are pretty good. (Kearney 1996)

These dual CEI functions were adaptations to the way the interactions between business and government, and among different government agencies and levels, had evolved over decades. The state labor and welfare agencies inhabited different worlds from each other and from the private sector, and economic development specialists had been frustrated by these divides for decades; CEI staff crossed the divides.

It may be that CEI had found a way to change the behavior of the firms and their owners. One CEI staffer thought so: "The firms are changing. They don't think of people as 'welfare' or 'AFDC' people so much, but as individuals with their own problems, but suited to their own individual slots. So they break up the stereotypes" (Stevens 1996).

Some in CEI might go so far as to suggest, in consequence, an emerging and different relationship between owners and labor. One factor that supports this view is the perception of professionals who train the "disabled." CEI's ETAGs in fact overlapped, in the firms they had located in, as much as 90% with the work of another organization, Employment Trust, Inc. (ETI) had begun as a training organization for severely disabled persons, but had spun off a subsidiary that supplied temporary workers to firms, including such a diverse spectrum of disabilities and support services that ETI was looking to expand to include welfare-to-work persons and others who were not disabled in any official sense. ETI found that it and CEI had discovered a niche in the Maine economy composed of firms that fervently believed in restructured labor relations as the key to increased productivity (Peterson 1997).

Training and placing this population in socially useful roles and jobs has created an opinion—perhaps not shared by a majority, but still much more prevalent than in the past—with the potential to change attitudes in other areas:

that all contributors are "differently abled," whether persons with physical or mental disabilities, those on AFDC for any of dozens of specific reasons, or the CEOs of companies. One could find this viewpoint among the staff at CEI, who claimed it existed in at least some of the companies that had experienced the ETAGs, as a result of the experience.

Conclusion

The overall question is, How do the have-nots—say, the lower part of the income and wealth distribution—achieve a place in the debate about economic policy? How do their ideas gain a permanent place in the thinking and policy making of the larger community? Two contrasting routes to such representation fell short in the period after 1981. One was the postwar pattern of labor representation based in industrial unions and operating nationally as a countervailing force toward corporate interests. This labor structure had become weakened and bureaucratic, its decline a big factor in the decline of liberal politics after the 1960s. Another route, also limited, was the path of community-based "grassroots" organizations. These had proliferated since the 1960s, but, despite notable exceptions, their negativism, fragility, and inability to form coalitions with each other or with labor had held them back from a permanent and significant place at the political table.

The possibility that occurred to some after 1980 was that labor and community interests might join forces. This began to happen in Maine and throughout the nation in the mid-1990s. Moreover, what we see are not just "community-labor coalitions" in a general sense, but a set of more specific organizational developments that gave them force—the emergence of an economic and community base for politics, rather than simply lobbying at the state house.

One could see such a base in the efforts of Coastal Enterprises, several women's organizations, the labor buyout capacity, and the advocates of the HPWOs, all organized around the goal of creating "good jobs." This constellation of organizations and lobbying energy shared several features. First, they had developed a critical mass over several years. CEI had been executing ETAGs since 1983, was providing business counseling for roughly 1,000 firms, and monitored training for perhaps 1,000 persons annually, at least half of them workers who, through welfare status or disability, would not otherwise have been in the labor force. MCWWC had been servicing upward of 1,000 women each year for two decades. The Hathaway and Biddeford buyouts represented a new role for labor, whose mass was still a significant part of the state's labor force. The firms to which Hillard and Vail referred, although only a small minority, still represented a cutting edge in terms of productivity. Moreover, these organizations had weight not only in the state legislature, but also in the deliberations of the private sector and nonprofit worlds.

Over time, these organizations had discovered not simply a new way to treat employees, but how to put an organizational force in place within firms and unions. The ETAGs involved both advocates within small manufacturing firms and supports outside the firms, tied together contractually. The HPWOs in larger workplaces found that alliances between labor union locals and certain management advocates were the key persuasive forces. The labor-led buyouts created patterns of teamwork involving new investors, community economic development officials, state agencies, and business management people. The women's organizations created a statewide network that monitored management practices in every part of the economy.

Second, a broadly coherent doctrine united these efforts. The new initiatives, on the one hand, seemed to be adaptations to new realities. The "high-performance workplace organization" noted by Hillard and Vail was, arguably, the best way to adapt to a world where "low performance" could not be sustained in the face of lower-wage nations. The employee buyouts were economically and socially justifiable as filling a niche left open by hypermobile capital and occasionally obtuse and overcentralized financial management. The programs of MCWWC filled a niche created by the dissolving ties of the traditional family, making it possible for women to obtain economic independence, while the downsizing and contracting out by larger firms created small business opportunities. CEI's ETAGs were an organizational innovation that adapted to the impossibility of a state agency monitoring increasingly complex relationships among training, finance, and firm-level productivity issues.

One could say that these innovations might not count for much if the counteracting pressures were strong and cohesive. Traditional labor relations and entrepreneurial forms were, however, not as strong as they had been. What had worked, or seemed to work, in the past did not have the same force when posed against some of the newer ideas. The state human services and economic development agencies, then, had to be defensive about their policies. The rationale for these government agencies derived from a New Deal-era rationale that had been defeated by the ideological and antibureaucratic pressures of the previous two decades.

A third factor was the emergence of a set of institutions backing the positions that had been perceived at the grassroots level. That the Maine Equal Justice Coalition had been able to get first-class work from Stephanie Seguino to establish a revised "living wage" threshold, and that the Maine Center for

Economic Policy could mobilize Hillard and Vail to write a persuasive gloss on the high-performance workplace, suggests an intellectual resource to counter what had long supported traditional policies. These pieces were, in fact, the tip of the iceberg of research work that had been going on for a long time, and with which policy advocates and legislators had become familiar.

All this added up to more than either a "liberal resurgence" or, in contrast, a "grassroots movement" alone. Maine fit neither pattern; it was building new institutions. The pieces came together in a political program that of itself was only the surface manifestation, and in any event far from a "victory" for the good jobs coalition. The coalition shared frustrations: The governor was mainly interested in satisfying a business constituency's need for profits, and was averse to direct intervention in the way firms deal with their employees. The Human Services commissioner, facing the challenge of putting AFDC mothers to work, was announcing that "any job is better than no job," while FAME was maintaining its disinclination to put conditions on subsidies; both positions undercut the jobs coalition program.

There was no clear victory, but the good jobs coalition was clearly at the table with other, longer-standing interests, and could hope for a period of compromise and incremental success. One had the impression of progress, labor and community groups were making connections, the climate of tolerance seemed strong, and whatever setbacks occurred, the good jobs forces would be in the field for years to come.

Notes

1. Aside from the practical obstacles to community-labor alliances, and underlying them, are profound differences in doctrine. Twentieth-century "liberal" thought has tended to rely on the primacy of national coalitions of "have-nots," primarily given voice through the role of organized labor, and distrusted localist aggregations as inherently conservative. Those who have argued for grassroots organizing have tended to distrust electoral politics and national support groups as co-optation (Piven and Cloward 1977).

2. This characterization of the good jobs movement is based on a review of newspaper and other articles done through a Nexus search, supplemented by telephone interviews. The count of "at least 24 states" is ours. Key references are for Baltimore (Peschek 1997; Cooper 1997), New York City (Polner 1996), Minneapolis (Diaz 1997), Los Angeles (Manzillas 1997), Houston (Burtman 1997), New Orleans (Ruggless 1997), Albuquerque (Henderson 1997), Florida and Seattle (Burney 1997), Portland and Gary (Rubenstein 1996), Milwaukee (Norman 1996), and upstate New York (Luster 1997).

3. The general background on Maine comes from a series of interviews and site visits to ETAG programs in factories during April, May, and December 1996. Some of this is contained in Clavel (1996).

4. Michael Hillard (1997) quoted a state economist, Laura LaChance: Jobs lost during the 1990-1997 period had average annual salaries of $29,000, whereas the average salary for jobs gained was $19,000. And on rural versus urban disparities: The state average versus the southernmost York and Cumberland Counties—unemployment was 2% higher, the poverty rate 50% higher, and per capita income 24% lower.

5. Kit St. John (1997) held the opinion that DHS Secretary Concannon "made a political judgment that the public simply wanted to put welfare clients to work. . . . Neither the commissioner nor the deputy commissioner have an economic policy. They are driven by their perception of the politics, that people have decided welfare recipients should go to work. The rest is rhetoric."

6. Rick Dieffenbach (1997), Maine Quality Center director, noted an increase in volunteer examiners from 30 to 78, drawn mainly from firms that supported the quality jobs idea. Vail also noted that the speakers had all said success in creating the HPWO involved a multiphased progression, yet all speakers showed that success was accomplished a step at a time. Vail's coauthor, Michael Hillard, told us: "We need to both respect the tradition of self-employment and small firms and also realize that we can move beyond that."

7. In 1996 there were 75 active ETAGs, and there had been a cumulative total of 140. These had resulted in the creation of some 1,400 jobs for low-income persons, somewhat less than half the cumulative expansion of jobs through the program.

References

Armstrong, D., and E. O'Brien, 1997. Without jobs that pay a living wage, little will change for the struggling families of rural Massachusetts. *Boston Globe,* March 11, metro, A1.

Barlett, D. L., and J. B. Steele. 1992. *America: What went wrong?* Kansas City, MO: Andrews & McMeel.

Beaudoin, J. 1997. Employees to share ownership of factory. *Portland Sunday Telegram,* March 28.

Blom, E., and A. Garber. 1996. Rich and poor: The growing gap. *Press Herald* and *Maine Sunday Telegram,* April 29-May 5.

Brigham, A. 1997. Remarks made at the "Taking the High Road . . . " Conference of the Maine Development Foundation, University of Maine, Augusta, May 7.

Bull's-eye of success [editorial]. 1996. *Bangor Daily News,* January 18.

Burney, T. 1997. Grass-roots persuaders cultivate decency on job. *St. Petersburg, Florida Times,* March 9.

Burtman, B. 1997. Minimal theories: Despite the rhetoric, evidence on the minimum wage initiative isn't conclusive either way. *Houston Press,* January 16.

Cavanaugh, M. 1997. Interview. May 19.

Clavel, P. 1996. *Community development in Maine: Coastal Enterprises, Inc.* Working Papers in Planning 155. Ithaca, NY: Department of City and Regional Planning, Cornell University.

Cooper, M. 1997. When push comes to shove: Who is welfare reform really helping? *The Nation* 264(21):11-15.

Diaz, K. 1997. Minneapolis City Council OK's resolution on "living wage." *Minneapolis Star Tribune,* March 8.

Dieffenbach, R. 1997. Interview. May 16.

Fitzgerald, D. 1997. Remarks made at the "Taking the High Road . . . " Conference of the Maine Development Foundation, University of Maine, Augusta, May 7.

Halper, E. 1996. The TIF race is on: State likes giving taxes back to business, but others aren't so sure. *Maine Times,* May 9.

Hastedt, C. 1996. Interview. December 16.

Hillard, M. 1997. Opening statement made at the "Taking the High Road . . . " Conference of the Maine Development Foundation, University of Maine, Augusta, May 7.

James, F. 1995. Baltimore tries out "living wage" requirements: Goal is to lift low paid workers above poverty. *Chicago Tribune,* December 28.

JOBS NOW Coalition. 1995. *The job gap study: Is Minnesota creating enough jobs that pay a livable wage? Phase two: Beyond the numbers.* St. Paul: JOBS NOW Coalition.

Katznelson, I. 1981. *City trenches: Urban politics and the patterning of class in the United States.* New York: Pantheon.

Kearney, K. 1996. Interview. May 7.

Luster, M. 1997. Interview. Ithaca, NY, January.

Macpherson, C. B. 1977. *The life and times of liberal democracy.* New York: Oxford University Press.

Maine Development Foundation. 1996a. *Measures of growth.* Second report of the Maine Economic Growth Council. Augusta: Maine Development Foundation.

Maine Development Foundation. 1996b. *Measures of growth 1997.* Third report of the Maine Economic Growth Council. Augusta: Maine Development Foundation.

Maine State Legislature. 1996. *Report of the Commission to Study Poverty Among Working Parents.* Augusta: Maine State Legislature.

Manzillas, J. 1997. Telephone interview with California State Senator Tom Hayden, Los Angeles, May 27.

Nardone, G. 1997. Interview. May 9.

Norman, J. 1996. Despite criticism businesses renew vow to jobs program. *Milwaukee Journal Sentinel,* July 26.

Peschek, J. 1997. A living wage? Campaigns attach strings to public contracts. *Dollars and Sense,* March-April, 28-29, 34.

Peterson, R. 1997. Interview. May 7.

Pingree, R. 1996. Interview. December 16.

Pitroff, M. 1996. Living wage campaign working to replace minimum wage. Transcript 1852-9. *Morning Edition,* National Public Radio, April 23.

Piven, F. F., and R. Cloward. 1977. *Poor people's movements: How they succeed, why they fail.* New York: Pantheon.

Polner, R. 1996. The fear of wages. *City Limits,* February.

Porter, J. 1996. King rivets attention on business growth. *Maine Sunday Telegram,* January 14.

Rubenstein, S. 1996. Jobs with justice. *Oregonian,* February 8.

Ruggless, R. 1997. Minimum wage hike bills springing up nationwide. *Nations Restaurant News,* February 3.

Sandler, L. 1996. Light rail gets friendly reception. *Milwaukee Journal Sentinel,* December 11.

Seguino, S. 1995. *Living on the edge: Women working and providing for families in the Maine economy, 1979-93.* Orono, ME: Margaret Chase Smith Center for Public Policy.

Stevens, A. 1996. Interview. May 16.
St. John, C. 1996. Interview. December 12.
———. 1997. Interview. May 12.
Vail, D. 1997. Remarks made at the "Taking the High Road . . . " Conference of the Maine Development Foundation, University of Maine, Augusta, May 7.
Vail, D., and M. Hillard. 1995. One cheer for Governor King's economic development strategy. *Maine Center for Economic Policy News* 2(1).
Will, G. 1997. Tom Hayden's small vision for a big city. *San Francisco Chronicle,* February 10.
Williamson, B. 1996. King has been bamboozled. National Semiconductor reaps tax break, job cuts. *Morning Sentinel,* June 28.

12

Changing the Constraints

A Successful Employment and Training Strategy

Brenda A. Lautsch
Paul Osterman

Over time, a wide array of training approaches have been embraced by politicians hoping to improve labor market outcomes for those who need assistance. From the multiple programs established in 1962 with the Manpower Development Training Act, through consolidation under the Comprehensive Employment and Training Act (CETA) in 1973 and a shift to federal control under the Job Training Partnership Act (JTPA) in 1983, politicians have worked to come up with the right recipe to allow disadvantaged participants in the labor market to improve their human capital attainments and reap greater rewards. Policy analysts have contributed to this effort by developing and refining evaluation tools to measure program operation, shifting over time from a reliance on quasi-experimental to more desirable random-assignment evaluation methods.

Despite these efforts, there remain three related and difficult problems. First, it is not clear that public training programs are as effective as their advocates hope. As three recent reviews of the evaluation literature make clear, over the years, training programs have proven to be worthwhile social investments in the sense that benefits exceed costs. Nonetheless, very few, if any, have achieved gains that are large enough to propel their clients out of poverty and onto a substantially higher earnings trajectory (U.S. Department of Labor, 1994; LaLonde 1995; Grubb 1996a). Second, we have as yet little insight into the mechanisms of program success. Traditional experimental evaluation approaches are well suited to capturing mean program impacts, but less appropriate for entering the "black box" of program operation. This gap in the evaluation literature has left us with little scholarly insight into how to improve or redesign programs. Finally, the economic and policy environment in which training is occurring is changing. Training programs themselves are evolving in response, and the character of these changes makes more critical a new approach to assessing them. In particular, low-skill, high-paying jobs are disappearing, product and labor markets are becoming more volatile, and policy is being devolved to the state level.

Connected to the last of these developments, new types of training programs that are more attentive to the local environment and involved in reshaping it are becoming increasingly common. So-called sectoral programs attempt to combine employment and training with interventions among firms (Aspen Institute 1995). These interventions are intended to alter the hiring and job design patterns of firms and hence act upon the environment as well as upon individuals. One well-known example of such a program is Focus Hope in Detroit, which is aimed at the machine tool industry. Another related model is the Center for Employment and Training in California, which has strong ties with employers and a grounding in a community-based organization. In these and similar programs, evaluation cannot be limited to a traditional experimental approach of identifying what works and what does not for the average participant; the scope of intended impacts is much broader than this. In this chapter, we provide an example of a nonexperimental evaluation of this new type of training. We report on a program, Project QUEST in San Antonio, Texas, that we think is a model of how a successful employment and training program can change the constraints facing labor market participants.

QUEST has been successful for reasons related to its structure—and not, for example, because it is managed by a charismatic leader—and this structure provides lessons for program design. The three most important characteristics of the program, which we believe explain its success, are the quality of its relationships with employers, its interaction with the community college system, and its base in a strong community organization. These three characteristics are interesting on their own terms. However, when considered as a group they add up to something more: What distinguishes QUEST are its efforts to alter the environment within which it and its clients operate. This larger ambition represents an approach toward training policy that we think holds considerable promise. Indeed, a larger conception of the role and possibilities of employment and training programs points to a strategy with more promise than past efforts.

The Research Design

The research reported here was commissioned by the Ford Foundation as an assessment of Project QUEST. We visited San Antonio repeatedly over the course of more than a year and employed a three-pronged research strategy. We conducted more than 65 face-to-face interviews with a wide range of actors (clients, program administrators, employers, local educational officials, community activists, and public officials) as well as three focus groups with clients. We administered a telephone survey directed to all persons who had participated in the program as clients. Finally, we also had access to the management information system (MIS) maintained by Project QUEST, which contained considerable information on clients.

As a result of these efforts, we obtained very rich information on the structure and operation of the program as well as its institutional impacts in the local community. We also obtained considerable information on the characteristics of the program's clients as well as pre/post outcome data on them. However, we entered the scene long after it was possible to implement a random-assignment evaluation. Some readers may find this an insurmountable obstacle to taking our results seriously. We would make several points in response. First, some of the important issues raised by Project QUEST center on institutional changes for which random assignment is irrelevant and that are best addressed through some of the methods we employ. Second, as will emerge below, the impacts of the program appear to be so large and so linked to the program mechanisms that it is very hard to believe that they would be overturned by a more rigorous evaluation. In addition, we are able to address some aspects of selection concerns with the data that are available. More broadly, we are not interested in this chapter in providing precise point estimates of impact or engaging in cost-benefit analysis. Rather, we want simply to establish the success of the program in impact terms and then go on to discuss the broader institutional reasons that lie behind that success. The impact estimates we present are so large that we think that this strategy makes sense.

A Description of Project QUEST

Most accounts date the momentum for the development of Project QUEST from the January 1990 announcement of the closing of a Levi's plant in San Antonio that employed more than 1,000 people. In response, two community-based organizations, Communities Organized for Public Service (COPS) and Metro Alliance, began to organize around jobs and training. Both organizations are affiliates of Industrial Area Foundations (IAFs), a national network of community-based organizations that dates back to Saul Alinsky's mobilization of "people's organizations" in the poor neighborhoods of Chicago in 1940.

In neighborhood meetings, people told stories of past unsatisfactory experiences with training programs, particularly those from which no recognized certificates or diplomas resulted, or in which there were no available jobs after the training period. Through discussion of such stories, three principles were developed that would guide Project QUEST: First, training must be long-term; second, training must be job driven because, as one IAF leader stated, "our people had enough disappointments in their lives"; and third, there must be adequate individual financial support.[1]

An important part of the efforts of IAF leaders involved generating support for the training initiative from the San Antonio business community. They started by gaining the involvement of a few influential employers and then leveraged their commitment in larger meetings of business leaders. Ultimately, a large number of employers expressed commitment, promising to provide information on their evolving staffing requirements, to cooperate in curriculum development, and to offer financial or other support.

Program Design

Project QUEST started enrolling participants in January 1993. Between that time and December 1995, when we closed the study, a total of 825 people had participated in the program. Of these 825 people, 629 had left the program, and of those who had left, 447 were positive and 182 were negative terminations.[2]

Those eligible for training through QUEST are chosen through a two-stage procedure. In the first stage, called *community outreach,* applicants are interviewed by IAF leaders in a local church or other community center. The outreach workers collect preliminary information about applicants and fill in intake forms. They also provide applicants with information about the occupations that are available for training and about the degree of commitment required from

QUEST students. Following community outreach is an intake procedure at the QUEST office. Staff members at this point exclude from participation in QUEST any applicants who have not attained a high school diploma or general equivalency diploma (GED). Applicants must also be economically disadvantaged and must have at least one additional barrier to employment (for example, be a single parent or have an arrest record). The exact income cutoff depends on the funding source supporting the specific student. QUEST staff also test applicants for the minimum required math and reading competence and for occupational aptitudes.

The training that Project QUEST provides is distinctive in three ways: in the involvement of employers in the program, in its reliance on community colleges, and in the amount and nature of support given to participants. Project QUEST began with commitments from employers to provide jobs to 650 participants. These up-front commitments were intended to distinguish the program from others that trained without knowing where the placements would come from; we will discuss the nature of these commitments below. An additional element of the program model involved working with employers to design training curricula and to forecast future needs. QUEST staff also worked closely with the community colleges in designing appropriate and innovative curricula. Training took place almost exclusively in these colleges in a number of different occupational fields, but with a notable focus on medical occupations. The distribution of students across these fields is shown in Table 12.1.

Project QUEST is further distinguished by the extensive counseling and financial support it provides to participants. QUEST participants receive assistance from counselors in navigating the community college system and with a wide range of emotional, personal, and family support. Approximately 90% of counselor time is allocated to client-related activities, with the remainder required for program administration. Of this 90%, about half is academic. This involves speaking to instructors, tutoring, and keeping in touch with what's happening in school. It also includes motivational group sessions called VIP meetings (for Vision, Initiative, and Persistence). These meetings are held once a week, and attendance is mandatory for QUEST participants. In these sessions, counselors keep students updated on program information, such as when new ID cards must be made and when forms must be completed. The VIP meetings also provide a forum in which students can air their concerns and a way for counselors to track student progress in class.

The remainder of counselor time is split between emotional support and financial support services. Emotional support involves talking things out with students and providing referrals to other public or community resources. Examples of the issues that counselors may deal with include spousal abuse, student

TABLE 12.1 QUEST Occupations

Training Field	QUEST Students Enrolled	Percentage of QUEST Students Enrolled
Licensed vocational nurse (LVN)	165	20
Registered nurse (RN)	57	6.9
Medical technician of varied specialties	233	28.2
Health unit coordinator	29	3.5
Medical assistant	13	1.6
Social services technician	8	1.0
Child-care technician	5	.6
Customer service representative	18	2.2
Office systems technician	57	6.9
Financial services provider	53	6.4
Accounting technician	14	1.7
Legal assistant	5	.6
Network administrator	16	1.9
Computer programmer	7	.9
Facilities technician	10	1.2
Aircraft maintenance worker	15	1.8
Plumber	10	1.2
Electrical technician	36	4.4
Diesel mechanic	37	4.5
Miscellaneous	37	4.5

low self-esteem, and parenting challenges. A final and critical part of the counselor's role is to assist economically disadvantaged QUEST participants in accessing available financial support. Counselors provide advice to participants on how to budget their resources, provide references to other private and public aid that is available, and administer QUEST financial support.

This financial support is "categorical" in that it covers specific items—transportation, child care, rent, eyeglasses, and clothes—rather than being a general stipend. Hence different individuals receive different amounts. The average monthly support payment is $240.

Two points are important regarding financial aid. The first is simply that it exists—one of the most significant shifts involved in moving from CETA to JTPA was the elimination of stipends. The rationale was to eliminate the welfarelike aspect of CETA programs, but this had the consequence of limiting interventions to short time periods. By drawing upon multiple sources of funding, Project QUEST has opted for a different model. Second, the stipends, although generous by JTPA standards, nonetheless are not enough to live on and require that participants hold part-time jobs, receive help from family members, or remain on welfare.

Program Impacts

In this section we provide a profile of Project QUEST clients and the outcomes of the program for them. Elsewhere we have documented these program impacts in more detail (Osterman and Lautsch 1997), so here we briefly present the size and direction of effects and focus more on the mechanisms underlying them.

Our estimates of program outcomes draw upon data from two sources: Project QUEST's management information system and a telephone survey. The Project QUEST MIS contains information for all 825 participants on demographic characteristics, some of the employment barriers they faced, on their pre-QUEST employment experience, and (for positive terminations only) on wages and other characteristics of the jobs they obtained after leaving QUEST.

We conducted a survey by telephone of 541 of the participants, including those still in the program as well as positive and negative terminators. The survey collected information on the experiences of the participants while in QUEST, on some aspects of their pre-QUEST experiences, and on what happened to them after they left Project QUEST. Because this survey included negative as well as positive terminations, we will use the survey results in describing economic program effects.[3]

Table 12.2 provides descriptive statistics on the characteristics of Project QUEST participants. As is true of most programs of this type, the participants are mostly women and minority group members. However, unlike in most welfare-to-work programs, a substantial minority of the participants are men. The education level of the participants is high (a high school diploma was an admission requirement), but it is important to note that more than 30% of the high school diplomas are in the form of GEDs, which are typically seen as lower quality (Cameron and Heckman 1993; Murnane, Willett, and Boudett 1995). Finally, about half of the participants were receiving welfare and/or food stamps at the time they entered the program. We will provide preprogram employment and earnings data momentarily, but it is worth noting here that less than half of the participants were working at the time they applied to the program, and their average pay was quite low.

Table 12.3 displays information on the pre- and postprogram employment of participants. The preprogram employment rates refer to the time when participants applied to QUEST, whereas the wage and hours data are for the most recent

TABLE 12.2 Characteristics of QUEST Participants (in percentages)

Female	65
White	18
Black	11
Hispanic	69
Other	2
Regular high school graduate	70.1
GED	29.1
Some college credits	45.3
Receiving welfare at time of admission to QUEST	44.5

jobs they held prior to QUEST entry, which can include jobs they held at the time they entered or may refer to jobs from several years before. The post-QUEST employment data refer to the time of our survey, which can vary from a month to about a year and a half since leaving the program. The regression reported below controls for all of these timing considerations. The data represent all program leavers, positive as well as negative terminations, and all monetary figures are in 1995 dollars.[4]

It is apparent that the clients of Project QUEST experienced substantial gains in terms of probability of working and wages and hours conditional upon working. If we multiply the probability of working times hours worked times hourly wage times 50 weeks, then the average QUEST participant (positive and negative terminators taken together) gained $7,457 over his or her pre-QUEST annual earnings. Taken at face value, this improvement suggests increases in annual earnings that are very far in excess of the typical employment and training results. We will take up below some qualifications and cautions regarding these

TABLE 12.3 Pre/Post Economic Comparison

Percentage working at time of QUEST admission	48.5
Percentage working after leaving QUEST	73.0
Hourly wage at most recent pre-QUEST job	$5.99
Hourly wage at post-QUEST job	$8.41
Hours per week at pre-QUEST job	32.4
Hours per week at post-QUEST job	39.3

NOTE: The data in this table are limited to QUEST participants who had left the program as of December 1995. All wages are in 1995 dollars. Pre-QUEST employment data are from the MIS; post-QUEST data are from our survey. The pre-QUEST wage and hours are for the job held most recently prior to QUEST entry, and in some instances people continued to hold these jobs while in QUEST. Individuals ($n = 88$) who had never worked prior to QUEST are not included in the pre-QUEST wage and hour calculations.

outcomes, but we will also argue that these impacts are so large that they are very unlikely to be overturned. It is worth noting that, according to the Texas Employment Commission, the average hourly earnings of manufacturing employees in San Antonio rose 14% between September 1992 and December 1995. To obtain data on all employees, not just in manufacturing, we need to look at Bexar County, and for these employees between the third quarter of 1992 and the second quarter of 1995, wages rose 8%. Both sets of earnings gains are clearly well below the increase experienced by QUEST participants.

We should also note that these results show that the payback period for QUEST is very short. Although the program is expensive, it appears to be a good investment. The average cost per participant is $10,077. Of this cost, 56.1% goes to direct payment for the participant (tuition, child care, food, books, and transportation), 23.1% goes to support staff (counselors and occupational analysts), and indirect costs account for 20.7%. The total cost for a positive terminator (who averages 17.7 months in the program) is $11,066. Looking purely at program costs and employing our most conservative estimates of earnings gains, Project QUEST returns 100% of its investment in 2 years.

Threats to Validity

One concern that might arise is that the selection process itself produced biases, specifically that the preapplication screening that was conducted by the community groups led to the exclusion of less motivated or less capable people. In fact, the purpose of the community screening process was more political than programmatic. The community groups had expended considerable political resources establishing QUEST and were concerned that participants and their families understand from whence the program came. Hence the initial screening was designed to establish a link between the groups and the participants, and in fact this proved helpful later when the community groups became more active in the effort to reduce dropouts. In order to avoid political problems (e.g., by seeming to favor some persons in their constituencies over others), the interviewers were under specific instructions not to accept or reject people, but rather simply to explain the program and establish a relationship. This point was made to us clearly in a number of interviews. It is no doubt true that when some people found out the program involved a commitment of nearly 2 years they withdrew, but this would have happened anyway and is not properly interpreted as creaming.

A second concern, fueled by the relatively high educational attainment of QUEST participants, is whether they would have made it on their own without the program. As a preliminary point, it should be recognized that the program in fact aimed at a group of people who were not at the very bottom of the labor market—that is, who had at least a high school diploma or the equivalent. Hence the QUEST participants were not the most disadvantaged people one could find, but this in itself does not mean that they did not in fact face significant barriers and would have done as well without help. We have gathered several pieces of evidence that suggest that QUEST made a real difference.

First, we randomly selected 35 case files from the QUEST records and read these files in detail, paying particular attention to the counseling notes. Our goal was to see if these persons seemed to face significant barriers that would make it very difficult for them to do well in the labor market on their own. Not surprisingly, all of the 35 people faced the problems of poor work histories and low test scores. Beyond this, however, 12 had serious health problems, 8 had significant family disruptions (abusive partners, children continually in trouble, and so on), 3 had personal difficulties with alcohol, and 3 had problems with the criminal justice system. In each case the files recorded the actions taken by QUEST and its counselors to intervene. It seems likely to us that these actions by QUEST played a significant role in enabling these people to make it through the program and into jobs.

Second, our interviews with employers and with community colleges provide concrete evidence of the mechanisms through which QUEST made a difference. In the higher-education system we interviewed senior administrators, program directors, and teaching staff. Their perceptions were quite uniform. For most programs, they said that QUEST students had the same demographic profile as the typical community college student but that QUEST students arrived with weaker academic preparation. Virtually all indicated that the QUEST students had lower dropout rates than other students, and they attributed this not to any greater motivation on the part of QUEST students but rather to the constant presence of QUEST counselors on the campus.

The employer interviews are even more persuasive. The hospitals indicated that the demographic profile of QUEST students in registered nurse jobs is different (more disadvantaged) from that of non-QUEST students. This implies that QUEST students would have been unlikely to land these jobs before their participation in the program. An interview at a bank suggested that QUEST students have more personal problems than typical hires but that the counseling provided by QUEST makes a positive difference. Until QUEST, employers who wanted diesel mechanics hired exclusively from a technical school in Waco. QUEST revived and made credible the diesel program at a local community college, and thus opened up a well-paid occupation that was simply unavailable before.

The third piece of evidence concerns creaming. Elsewhere, we have reported on a detailed examination of the creaming issue in which we asked who among QUEST participants gains more from the program than others (Osterman and Lautsch 1997). The data we collected contain considerable information on characteristics of QUEST participants. If creaming were important, we would expect that individuals with advantages—such as having a regular high school diploma instead of a GED—would do better than others. Instead, we found few significant relationships between individual characteristics and various measures of program outcomes. QUEST was able to succeed with a wide range of people. In this sense, concerns about creaming do not appear to be of central importance. We also explored the so-called Ashenfelter dip (Ashenfelter 1978), and though we found some evidence of this pattern, it was not sufficient to explain away our findings.[5]

In summary, we grant that the average impact of QUEST cannot be assessed with the confidence that would be possible using random-assignment evaluation. However, when we consider the array of evidence before us—the size of the impacts, our examination of individual participants' files, the testimony of employers and community college staff, the regressions—it nonetheless seems to us that we can have enough confidence in the results to take this program's lessons very seriously. We turn now to some of the mechanisms underlying these results.

Institutional Impacts and Mechanisms

Project QUEST's founders think of the program in more ambitious terms than as a traditional employment and training program. In a traditional program, the nature of the external environment—the behavior of firms, the surrounding educational institutions, and the community itself—is taken as a given, and the program simply seeks to place clients successfully in that environment. Project QUEST tried to become an active actor in the San Antonio labor market and education system and, by doing so, to encourage institutional change. If successful, this would mean that the impacts of Project QUEST extend beyond the experiences of the clients themselves. Therefore, in addition to the economic gains to clients, a full assessment of the effects of Project QUEST would include less tangible but possibly even more important improvements in the functioning of the San Antonio educational system and labor market. The implications of the fact that QUEST seeks to interact with and alter its environment lie at the heart of our analysis of what makes the lessons of this program important.

Employers

Project QUEST works with employers to identify likely future needs and to design training curricula to meet those needs. This extended interaction with employers has also led to some changes in employer behavior. In our interviews, we observed a number of examples. In one, QUEST worked with banks and with community colleges to develop a financial customer services training program. This was an occupational category and certification that had not previously existed, and QUEST identified the need and convinced the parties to combine some skills and create the new certificate. Also in banking, one of the bank presidents we interviewed reported that QUEST was instrumental in creating a forum in which bank presidents and human resources staff from a variety of banks talk more with each other about labor market issues.

In health care, Project QUEST has generally placed students in nursing programs, which, due to state regulations, have fixed curricula. In some instances, however, QUEST has been active in working with employers to shape jobs. In an occupation called *health unit clerk,* training was redesigned to reflect employer input. Employers requested that certified nursing assistant training be included in a revised curriculum, as the demands of the unit clerk position are no longer just clerical. This change came about through discussions among employers, community college administrators, and QUEST staff. In addition to the change in content, QUEST pushed, with some success, for hospitals to raise the wages for this occupation.

Community Colleges

The community college system in San Antonio provides nearly all of the training to QUEST students. QUEST found that it could not simply enroll students into existing programs. In some cases appropriate programs did not exist, whereas in other instances new organizational arrangements had to be made to accommodate the needs of QUEST students. The development of strong working relationships with the community college system and the institutional changes that were achieved are among the most distinctive characteristics of Project QUEST. We present four notable examples below.

First, QUEST quickly learned that its students required more remedial work than expected. QUEST worked with the community colleges to establish the Basic Skills Academy, a remedial program that covers material that was not previously available to community college entrants and is organized to encourage more flexibility than the typical community college program. The community colleges were sufficiently pleased with this innovation to make it available to all of their students.

Second, in two certificate programs, computer information systems and office systems technician, QUEST worked with instructors to implement a pedagogical device—open entry/open exit instruction—that provides much greater flexibility to meet the complicated schedules and the personal needs of QUEST students. This involved the development of new curriculum materials and testing. The community college staff expressed considerable pleasure with this innovation. One of the community college faculty commented, "Project QUEST is always quoting something to us about new ideas for our program."

Third, Project QUEST was instrumental in encouraging the community colleges to begin conversations with the San Antonio Manufacturing Association and the Contractors Association regarding training needs of employers in the area. More generally, prior to Project QUEST the community colleges were not oriented toward working with the employer community, and Project QUEST has played a major role in changing the perspective of the community colleges.

Fourth, as we mentioned briefly above, QUEST restructured a diesel mechanics program that was slated to be eliminated. The experience of QUEST with diesel occupations provides the strongest example of QUEST's altering other training institutions and merits a more detailed examination. QUEST currently trains diesel mechanics students at the South West campus of St. Phillips College (SPC). In early 1993, this program was in trouble. It had low enrollment and was not generating graduates. The college had been doing an internal review of its programs and had targeted the diesel program for dismantling, recommending that the Texas higher-education board eliminate the diesel program, along with others that were not operating well. This board has a mandate to approve all programs and to enforce standards concerning the competencies of students, the number of graduates, and placement rates. In May 1993, the board refused to approve the program cuts and told SPC to straighten out the problems with the diesel program. The whole college was in a period of transition. A new chancellor was also appointed who was interested in change and open to new program improvement ideas.

At this time, QUEST was exploring job demands of employers and had identified an interest in diesel mechanics, in large part because of the passage of the North American Free Trade Agreement. QUEST approached the college to talk about the diesel program, indicating an interest in placing QUEST students in the program and making suggestions about improving its operation.

One of QUEST's occupational analysts played a central role in the changes that occurred in the diesel program. This is evident in a memo dated August 25, 1993, from the deputy chancellor of the community college district:

> I have made arrangements with Jack Salvadore of Project QUEST to allow us to use the services of [QUEST analyst] Rich White to assist us with the revision of the diesel program and the certification of the Airframe and Powerplant Program. We're grateful for their assistance.... I am hopeful that Mr. White will provide the catalyst for these program revisions, and I ask that you assist him and make him a part of our team.

With the occupational analyst's involvement, the college had embarked on a number of program improvements by September 1993: removal of excess equipment from the diesel bay, training needs analysis, curriculum revision, and acquisition of new equipment. In October, program and QUEST staff also started to rearrange the diesel bay, work on enhancing the lab, repair training aids, and design new mock-ups and training aids.

One important change that QUEST staff initiated was to improve the operation of the college's occupational advisory committee. White brought in employers with commitments to Project QUEST and joined them with long-standing members of the SPC advisory committee. The committee is now composed of the original college committee, staff, faculty, and QUEST counselors and employer representatives. Prior to these initiatives, the college's advisory committee for diesel mechanics met, but only for the minimum number of meetings and duration required by Alamo Community College District regulations. The advisory committee became much more active in curriculum review, but particularly in getting donations of equipment and scholarships. According to one college administrator, the advisory boards in the past typically operated with college staff transmitting information on current program operation to industry people, but without much input in the opposite direction. The involvement of QUEST employers, who are committed to employing graduates of the program, increases the motivation and activism of the committee.

The training program has also changed significantly. Based primarily on employer input, a new cycle of classes has been created for QUEST students that is now being opened up to non-QUEST students and spread to other programs. The course cycle is now four courses per semester, 8 weeks per semester, 4 days on and 1 day off per week, and a 9-month completion time for the certificate program. (Originally, the program had been three courses a semester, 16 weeks per semester, and a year completion time for the certificate.) QUEST had pushed for these changes, saying that employers thought the 12-month course was too long and included unnecessary subject matter. Employers wanted generic training in the program; they wanted to train their employees in specific skills themselves. QUEST also wanted students to be in classes all day, given that the program had to pay day-care costs for some of them.

The curriculum in this shortened program cycle has also been altered, eliminating some subjects that are not essential for a basic entry-level mechanic. For example, some courses intended more for agricultural application of mechanical concepts were eliminated, and an introduction to computers was added. An internship was also developed for the program, to give students practical experience. Each syllabus within the program was also examined and updated.

SPC administrators and instructors continued to work with QUEST staff on devising and implementing changes in the diesel program. Chancellor Ramsay supported the program and worked with QUEST on needed improvements. He also made one important shift in college operations by appointing Linda Rodriguez to be dean of the South West campus. Prior to her appointment, there had been no local management of this campus site. Rodriguez was also sympathetic to QUEST objectives and worked well with program administrators. The diesel instructors also participated in program improvements by periodically arranging for donations of equipment, such as materials with which to build engine stands.

This cooperation between the college and QUEST staff ultimately generated tangible shifts in local training practices. Enrollment in the diesel mechanic program increased from about 7 students in 1993 to 47 in the spring of 1995. Some of this shift came at the expense of other training institutions; employers and college administrators now claim that fewer companies send their employees to Waco, Texas, for diesel mechanic upgrade classes.

The experience of the college with Project QUEST also led to other, broader changes. One closely related development is that the success of the diesel program fostered the design of a hydraulic forklift program. The college has also started more high school/college vocational training partnerships, partly due to the positive reaction to the diesel program. For example, the Northside School District in San Antonio decided in 1995 that it could no longer afford to keep its program. As a result, the district loaned its vocational training equipment to the college and enrolled students in joint-credit community college courses. Other programs within the college are also under revision, and college administrators are trying to mimic some of the changes they enacted in the diesel program: They are trying to make advisory counsels more active, to seek out information from employers, and to generate more equipment donations. The following excerpt from one of our employer interviews summarizes these impacts:

> We got involved in QUEST because we were having trouble finding forklift mechanics. It used to be that people out of that program at SPC would come out and not know how to change a fuel filter. I never would have thought of contacting the college to make it change. The boss had suggested this previously but I didn't want anything to do with the college because they were producing such useless students. I did talk to them several times and the college didn't seem interested. I had toured the facilities and seen that they had nothing. The college didn't ask what our interest was or for our input. Now I am on the college advisory committee and have a good relationship. A few years ago I would have run off people from SPC. They didn't know what they were doing. I never would have hired them. The QUEST guys are now motivated and adaptable. They are exposed to all kinds of things. I have had nothing but good reports. They have mechanical ability.
>
> The curriculum is teaching them the right things. QUEST is now our recruiting source. We will probably hire the two interns we have now and expect to put on quite a few more people.

Community Institutions: Changes for COPS and Metro Alliance

The final element of institutional change of this program concerns the community organizations involved in its founding. In the survey and in our interviews and focus groups, we found that Project QUEST has a small but meaningful community-building effect. Our qualitative data show that community leaders are energized by the high visibility and success of Project QUEST. According to COPS and Metro Alliance, community leaders contributed 16,000 hours of time in the interviewing and selection process. It is also certainly true that Project QUEST has given COPS and Metro Alliance more credibility in the community and political systems of San Antonio. The training program has also translated into at least one tangible benefit for the IAF: Our interviews revealed that a number of new parishes have joined the Industrial Area Foundation, in large part due to the strength of its performance on job training. We also found a slight increase in participants' awareness and planned activity in the IAF.

Discussion

The foregoing material establishes a strong case for Project QUEST's success in areas where many previous employment and training programs have failed. QUEST has made substantial changes in people's lives, shifting them to a qualitatively different trajectory. In addition, it is possible to point to institutional changes in the San Antonio labor market and educational infrastructure that are due to QUEST and that will multiply its impacts. These changes in San Antonio's labor market institutions have also fed back to and had impacts on the outcomes for QUEST clients as well as for the broader population. The important question is what general lessons can be learned from Project QUEST that will enable planners to design more successful training interventions elsewhere.

Project QUEST clearly established relationships with employers—with the demand side of the market—that were considerably tighter than is typical. The program seems to address squarely the question, Where will the jobs come from? At the same time, it is important to recognize that the nature of the link between the program and employers is more shaded than might appear at first glance. Early in the history of Project QUEST, and perhaps even now, the best-known characteristic of the program was the claim that Project QUEST had received commitments for 650 jobs from San Antonio employers, and all that remained was to select and train candidates for these jobs. Indeed, the role of COPS and Metro Alliance in approaching the San Antonio business community and using these contacts as the basis to build the program is one of the distinctive aspects of the program.

In fact, what is unique about Project QUEST is not so much the role of prior commitments as it is the success of the project in involving employers in the design of training and the thinking through of future labor market opportunities. It is not hard to see that solid employment pledges per se are difficult to make. Given the length of the selection and training period, if Project QUEST were to work literally according to the model of an employer placing an "order" or making a "pledge" and QUEST selecting and training candidates, a 2-year or longer lag between pledge and hire would result. It is a rare employer that can make such a commitment. Indeed, Project QUEST recognizes the reality of job pledges; one staff member described the pledges to us as really "moral commitments."

What is distinctive about Project QUEST is that it works with employers to identify *likely* future needs and then design training curricula to meet those needs. When this works well, QUEST functions as an extension of the firms' human resources departments and adds value because QUEST can have a broader view of the labor market and of educational institutions than can any single employer. As we have noted, in our interviews we observed a number of examples of QUEST functioning in this way. In addition, the process of working with employers creates a set of relationships in which "moral commitments" have force.

Second, it is clear that extensive training is a central explanation of the program's achievements. At the same time, it would be incorrect to come away with the impression that the main lesson of QUEST is that more expenditures are necessary. There are other job training programs that make significant investments in their clients, such as the Job Corps, and some of these are successful. However, QUEST is distinct in that it can have a broader impact than these programs because of the role QUEST plays in the day-to-day functioning of the San Antonio labor market.

Our third conclusion is that QUEST has succeeded in part because it has built into its governance and oversight structure a community organization that is powerful enough to obtain resources and sophisticated enough to act as a check and balance without at the same time seeking to take over and use the program for purposes that are not part of the central mission.

An additional organizational issue concerns the role played by community colleges. As we have noted, virtually all QUEST training takes place in community colleges and, indeed, some of the most important institutional impacts of QUEST have been upon these institutions. What is important to note here is that the community colleges constitute a stable and professionalized training system (Grubb 1996b). QUEST has not attempted to design and manage its own training infrastructure. Given the impermanence of many training organizations and the uncertainty of funding that has characterized employment and training policy, the organizational stability inherent in the community college system seems to us to be an important explanation of Project QUEST's success.

We have shown that QUEST has had a broad impact on institutional change, particularly in the San Antonio educational community, and in this sense has been more successful than a typical program. However, the significance of the point seems to us to be broader: QUEST's broader ambition has positive implications for its capacity to be successful in its narrower mission of job training. This occurs because QUEST has become an important "player" in education circles in San Antonio. As a result of its broader ambitions, it has reach, contacts, and legitimacy that typical employment and training programs lack. This in turn means that QUEST is better positioned to bargain with the political community for resources and with the business community for jobs. This illustrates the general point—which comes up also when observers study European programs—that the more central an employment and training system is to the day-to-day functioning of the labor market, the more successful it will be.

Notes

1. These principles were developed into the design of Project QUEST by Bob McPherson, an employment and training expert at the University of Texas, and Brian Deaton, now an employee of the U.S. Department of Labor.

2. In this chapter we refer to *positive* and *negative* terminations. These are JTPA definitions that we adopt for the sake of consistency with the management information system of Project QUEST. A positive termination is defined as one of the following outcomes: entered a job, entered the armed forces, entered a registered apprenticeship, entered non-JTPA training, completed a major level of education, or passed the Texas Academic Skills Program educational achievement test. A negative termination is a program departure for any other reason.

3. We designed the survey, and it was conducted by Galloway Research of San Antonio, under the supervision of Ester Smith of EGS Consulting. Extensive attempts were made to contact all 825 persons who had participated in Project QUEST up through the fall of 1995. The survey achieved a response rate of 66%. In testing for bias in response rates, we found that the survey obtained responses from a group that was more likely to currently be in the program than the sample as a whole and that was somewhat better off in terms of being weighted toward positive terminators and people with some college. Respondents were also slightly more likely to be female. However, the differences were not great. For example, according to the MIS, 23% of QUEST participants attended college and 54% were positive terminators, whereas in our survey 30% attended some college and 56% were positive terminators. Despite these differences, there is no significant divergence with regard to pre-QUEST earnings (which are taken from the MIS and hence are available for everyone). Among those who were captured in our survey, the average rate of pay in the job just prior to QUEST was $5.71 an hour, whereas among those we could not contact the average was $5.58. Looking two jobs back, the average was $6.14 per hour for both groups. As a further test, we ran a wage regression in which the dependent variable was the log of the pre-QUEST wage and the independent variables (also taken from the MIS) were age, sex, race, whether the person had a GED, whether the person had some college credits, whether the person had military experience, whether the person had an arrest record, whether the person was handicapped, and whether the person was a respondent in our survey. The survey variable had a small coefficient of .01 and standard error of .02, and hence was highly insignificant.

The foregoing analysis cannot rule out the possibility that people who failed to gain much from QUEST were harder to find for our survey than people who did well. However, based on observable characteristics, most persuasively their pre-QUEST earnings, there does not appear to be an important prior difference between the people we sampled and those we did not.

4. We used the U.S. Bureau of Labor Statistics' consumer price index for Dallas, Texas, because we were unable to obtain a consumer price index series for San Antonio.

5. This is the observation that people tend to enter training programs when confronted with a temporary decline in their fortunes. As a result, subsequent wage gains that might be attributed to program effects are really due to a return to the normal trajectory.

References

Ashenfelter, O. 1978. Estimating the effect of training programs on earnings. *Review of Economics and Statistics* 60(1):47-57.

Aspen Institute. 1995. *Sectoral employment programs*. Washington, DC: Aspen Institute.

Cameron, S. V., and J. Heckman. 1993. The non-equivalence of high school equivalents. *Journal of Labor Economics* 11(1):1-47.

Grubb, W. N. 1996a. *Learning to work: The case for reintegrating job training and education*. New York: Russell Sage Foundation.

Grubb, W. N. 1996b. *Working in the middle: Community colleges and the enhancement of the sub-baccalaureate labor force*. San Francisco: Jossey-Bass.

LaLonde, R. 1995. The promise of public sector sponsored training programs. *Journal of Economic Perspectives* 9(2):149-168.

Murnane, R., J. Willett, and K. P. Boudett. 1995. Do high school dropouts benefit from obtaining a GED? *Educational Evaluation and Policy Analysis* 17(2):133-148.

Osterman, P., and B. A. Lautsch. 1997. Lessons from success: A strategy for employment and training policy. Unpublished manuscript.

U.S. Department of Labor, Office of the Chief Economist. 1994. *What works, what doesn't*. Washington, DC: Goverment Printing Office.

13

Prospects for Job-Centered Economic Development

ROBERT P. GILOTH

Cleveland's Jobs and Workforce Initiative (JWFI) is an ambitious effort to reform workforce systems. It is regional and employer driven, broader than job-centered economic development in its universal approach to workforce development for all residents, yet it holds good jobs for the unemployed and underemployed as a central concern. Although it is too early to evaluate in terms of outcomes, the JWFI has attracted the Center for Employment Training to replicate in Cleveland with local providers, undertaken prototype training for customer service representatives with a Jobs Compact of 30 local providers, planned for regional one-stop centers, and advocated a state tax incentive for employer investment in training (Cleveland Growth Association 1997; Berry forthcoming).

The JWFI grew out of employer workforce concerns and the entrepreneurialism of a local civic infrastructure. At the same time there were 70,000 residents unemployed in the Cleveland region, one-third of 750 surveyed manufacturers stated that finding employees was a major need; a survey of 1,100 firms in multiple industries indicated that 44% had problems finding "qualified" employees. In 1996, the CEO organization Cleveland Tomorrow and the Cleveland and Gund Foundations joined with the Cleveland Growth Association and its 16,000 members to launch the JWFI.

Tight labor markets, national welfare and workforce legislative changes, and the ineffectiveness of many public systems are encouraging public and private innovations in local workforce planning, policy, and institutional design. This has created a policy environment that is complicated, evolving, and local. Maps are needed even as new pathways to work are being developed.

Workforce innovations take many forms. They include regional workforce boards, foundation-sponsored workforce intermediaries, livable wage coalitions, and state-level workforce planning. Oregon, for example, has developed a detailed set of benchmarks for workforce development that regional boards are slowly adopting. In the Silicon Valley and Louisville, employers and their partners have developed a variety of workforce innovations. Massachusetts has reinvented its Private Industry Councils into Workforce Boards. The congregation-based organizing group in San Antonio, COPS/Metro, has garnered business and state-level support for a new labor market intermediary, Project QUEST (Okigaki 1997). And multiple national foundations and employer associations are developing distinctive proposals for national labor force intermediaries that link with state and local efforts.[1]

As with the case of Cleveland, these efforts, although hopeful, raise many questions: Will low-income job seekers get lost in the race for institutional innovation? Will they be relegated to bad jobs without the supports to navigate an ever more complex labor market? How will learning and coherence emerge in this scene of devolution and local-level innovation? Will investment in long-term training still be possible? Again, what happens when the job market softens, or when the reality sinks in of too few good jobs for low-income people in central cities?

All these questions are of relevance for a consideration of the prospects of job-centered economic development as described in this book. There is a danger of losing the jobs focus if practitioners of job-centered economic development simply adapt to and pursue the opportunities of a changing policy environment. Alternatively, can this array of practices, strategies, and organizations grow itself as a field, or perhaps a movement, that can shape the trajectory and outcomes of current institutional innovation on behalf of good jobs for low-income communities? This will require theories about how to bring jobs projects to scale and strategies for navigating the scramble for workforce and economic development innovation.

This chapter explores these challenges from several perspectives. First, I review the major findings and insights from each section of this volume, summarizing areas of agreement and disagreement. Second, I reflect on the prospects for job-centered economic development given the emergency policy environment.

Reflecting on the Chapters

The contributors to this book focus on labor market strategies that have some track record in connecting low-income residents to good jobs, creatively address current labor market challenges and opportunities, and offer promising approaches for replication and policy reform. I have called these practices *job-centered economic development* in order to emphasize the importance of good jobs as an outcome of employment training, economic development, and welfare changes, the need for market-oriented and integrated strategies, and the discipline of results. This definition is a conceptual artifice, and I make no claim to its foundations except that it serves heuristic and advocacy purposes.

Although job-centered economic development is not new, it has recently matured and become more widely acknowledged. Like many community-based interventions, these jobs projects are complex, systems related, and emergent, making them difficult to describe and evaluate in traditional terms. Consequently, this book brings together practitioners, academic researchers, and policy makers to examine different aspects of job-centered economic development. Although for the most part they use the same language about labor market barriers and interventions, their voices, points of reference, and conceptual and policy lenses are quite diverse. Some draw attention to sectors and networks; others focus on work attitudes and community engagement. Few of the authors would necessarily endorse my definition of job-centered economic development; each has his or her own preferred definition. These differences hopefully have surfaced new questions and directions to challenge the jobs field.

What have we learned from this consideration of job-centered economic development? Four questions, I believe, help synthesize the contributions to this volume:

- What have we learned about contemporary labor markets?
- What have we learned about effective good jobs projects?
- What do we know about the replication of effective jobs projects?
- How can jobs projects be scaled up to shape public and private employment policies?

In addressing these questions, I pay particular attention to the insights of the chapter authors that aid the design of effective labor market interventions.

What Have We Learned About Contemporary Labor Markets?

Harrison and Weiss (Chapter 2) and Dresser and Rogers (Chapter 4) convincingly argue that contemporary labor markets are undergoing dramatic changes: creating jobs and lower wages, expanding the contingent workforce, and dismantling career ladders. Tremendous variation exists by region, industry, size of firm, gender, race, and ethnicity. Despite the jobs and poverty alleviation benefits of tight labor markets, many inner-city residents still face considerable barriers in seeking and obtaining good jobs.

Two implications of labor market changes are important for job-centered economic development. First, change and complexity in labor markets (products of structural change, globalization, technology, and flexible production methods) have increased the lack of clarity for both firms and job seekers. Lack of clarity, and its resulting uncertainty, encourages firms' subjective assessments in terms of defining skills or soft skills, hiring through internal networks, and choosing locations or investments. Creating low-wage jobs increases firm turnover and may lower productivity, yet there are incentives for firms to underinvest in human resources. For job seekers, lack of clarity means lack of information about pathways to careers, appropriate skills, and entry points into the labor market. Lower rates of unionization exacerbate this lack of clarity.

Second, labor market change creates conditions for labor market matching or intermediation that serves both firms and job seekers. For firms, intermediation must address economic concerns related to turnover, productivity, and expansion. For job seekers, intermediation must go beyond training to include job readiness, connection to employers, postplacement supports, and career development. Effective jobs projects address both sets of conditions.

Nevertheless, the magnitude of structural changes in labor markets—regionally, globally, and internal to firms—suggests that it will take time, incentives, and power to alter these labor market processes and outcomes for low-income communities. Better programs alone will not turn labor market inequality around.

What Have We Learned About Effective Good Jobs Projects?

Mueller and Schwartz (Chapter 3) remind us that many policies and programs, including training, welfare-to-work plans, enterprise zones, mobility strategies, and business incentives, fail to work for low-income communities. It is important to understand the meaning of "fail to work," because many interventions do have positive effects; in most cases, however, they are rather modest and short-lived, failing to get people out of poverty. Mueller and Schwartz argue that this failure is in part due to the narrowness of the interventions, and that more comprehensive and integrated approaches have more potential to alleviate poverty. Lautsch and Osterman's evaluation of Project QUEST (Chapter 12) and the evaluation literature on the Center for Employment Training support this conclusion.

Harrison and Weiss also affirm this conclusion in their preference for the term *workforce development* over *employment training*; the former includes the recruitment of firms and job seekers, assessment, training, matching with jobs, and postplacement work. Nittoli and Giloth (Chapter 9) report a historical example of workforce intervention comprehensive in both conception and action: the New Careers movement of paraprofessionals in the 1960s and 1970s, which included job creation, training, career ladders, and unionization. Dawson (Chapter 7) describes the work of Cooperative Home Care Associates (CHCA), a contemporary jobs project that creates jobs, provides training, enables ownership and asset building, designs career ladders, and integrates postplacement supports.

The contributors also identify three other dimensions of effective jobs projects. First, effective projects are targeted or customized to specific job seekers and employers—one size does not fit all. Clark and Kays (Chapter 6) discuss several techniques for differentiating job seekers and present examples of how jobs projects target their efforts to specific populations during planning and implementation. For example, Harrison's STRIVE story (Chapter 8) shows that a 3-week attitude adjustment, job readiness, and job search program worked for people who needed to become quickly connected to the labor market. Over time, however, STRIVE realized that it needed to add postplacement and skill development components if the program's graduates were to move ahead in the labor market. In San Antonio, the problem is underemployment, not getting a first job. Consequently, Project QUEST, evaluated by Lautsch and Osterman and shown to produce significant earnings gains, enrolls high school graduates who have some work experience. Completion of this program frequently takes up to 2 years.

Second, effective projects become deeply involved with firms, sectors, and clusters. In Dawson's language, they become "valued" actors. Bosworth (Chapter 5) provides a number of tools for practitioners to get beyond secondary data analysis about regional economies and actually begin talking with companies and their webs of competitors and suppliers about workforce and economic development issues. Dresser and Rogers and Nittoli and Giloth offer examples of how to engage in sectoral planning in a range of industries. Clavel and Westmont (Chapter 11) recount how Coastal Enterprises of Maine has made targeted employment agreements with hundreds of firms that receive access to affordable capital and business incentives. Project QUEST broke through the negative perceptions of employers regarding training programs, engaging them in curriculum and occupational design as well as securing job commitments from them. As the story of the Center for Employment Training demonstrates, effective jobs projects become "embedded" or networked with employers (Melendez 1996).

Finally, effective jobs projects have strong community ties. Dewar (Chapter 10) recounts two stories about communities and jobs—the stories of Baltimore's BUILD, and Minneapolis's Interfaith Action—demonstrating that engaging residents makes jobs projects more effective because citizens are stronger participants and advocates than clients, and residents have important assets for labor market interventions, such as social networks. One important aspect of STRIVE's success, as Harrison reports, is that the program's trainers have had experiences and awakenings similar to those of the participants—as ex-offenders, welfare moms, drug abusers, and neighborhood residents. They can make connections based on lived experience. Nittoli and Giloth report that effective paraprofessional or community outreach worker programs build upon the indigenous skills and experiences of community residents to deliver human services. Lautsch and Osterman found in their study of Project QUEST in San Antonio that COPS and Metro Alliance, congregation-based coalitions associated with the Industrial Areas Foundation (IAF), were instrumental in getting employers to the table, recruiting and supporting participants, and securing financial resources for long-term training in a policy. And COPS/Metro grew stronger because of its involvement with Project QUEST. STRIVE has recently joined with East Harlem Partnership for Change, another IAF affiliate, to strengthen its community accountability.

These are certainly critical elements for successful projects, but there are others that these chapters deal with only implicitly. Effective jobs projects depend upon social entrepreneurs, individually but also in the form of teams and organizations. How to identify and nurture this entrepreneurialism is a major challenge for job-centered economic development, particularly when we think about replication and scaling up. There is also the important issue of benchmarking the outcomes, quality, and costs of different jobs strategies.

What Do We Know About the Replication of Effective Jobs Projects?

One of the perennial stumbling blocks for social policy is that of replication—transplanting successful projects to new contexts and locations (Replication and Program Services 1993). As I have discussed job-centered economic development, replication seems even more challenging: Jobs projects are intimately connected to place, embedded in networks, and advocated and implemented by unique social entrepreneurs. I have noted the replication challenges facing the Center for Employment Training in Chapter 1 (Melendez 1996; Harrison and Weiss 1998). A current debate in the employment field, which I will discuss in more detail later in this chapter, concerns the role and need for national intermediaries to promote job-centered economic development and its replication, much like those intermediaries that exist in the affordable housing field (Stanfield 1997).

Two authors in this volume discuss approaches that contrast with traditional replication strategies. STRIVE takes the time in each new site to get to know the local context and key players, and how its job readiness model fits or doesn't fit. For example, the Chicago replication, according to Harrison, has toned down the "tough love" aspects of STRIVE, believing that the Chicago area's older participants, many of whom are long-term welfare recipients, are likely to exit the program when exposed to this kind of pressure and feedback. Yet other key STRIVE elements remain in place—postplacement support, for example. And the social entrepreneur who leads the Chicago effort, although out of the STRIVE network, is quite different—without a grounding in street-level experience. His trainers, however, have this experience and credibility.

Dawson describes the replication of Cooperative Home Care Associates in Boston and Philadelphia as being explicit about transferring what had been learned in the South Bronx while being open to new ideas. CHCA's key design elements are its mission to create quality jobs and care, core competencies related to training and supervision, an organizational culture of respect, and a worker ownership organizational structure. CHCA set up the Home Care Training Institute as its replication arm; it is staffed and financed to undertake replication strategically, rather than to attempt replication on an opportunistic basis. In a sense, CHCA's approach to replication draws upon the organization's own history, in which the Community Service Society of New York played the role of investing and nurturing CHCA through its early years and hard times. The replication staff identified states with favorable Medicaid reimbursements for home health care, found sponsoring community organizations, arranged financing, and identified and trained social entrepreneurs. The replication sites and CHCA have formed a network, one in which learning is increasingly moving from replication sites to the original project. Dawson describes this process of connecting a dynamic model to dynamic sites as requiring flexibility, training of new social entrepreneurs, attention to middle managers, and more involvement from organizational leaders than originally expected.

These replication examples embody three principles. The first is design replication—key elements of a successful project creatively packaged to fit new contexts. The second is an investor approach to replication—deploying the intellectual, social, and financial capital in a strategic fashion. The third is a learning network of replication sites that together grow collective knowledge about effective practice.

How Can Jobs Projects Be Scaled Up to Shape Public and Private Policies?

Examples of job-centered economic development are rather small and, in the past at least, have usually existed in spite of public systems. They often exist in parallel to, for example, community colleges. These projects occupy unique niches, serving 2 to 300 people per site at best, and even replication does not enable them to reach scale in any one place. Jobs projects cost from $1,500 to $10,000 or more per placement, depending upon the nature of training and postplacement support. Although several well-known examples of job-centered economic development have received rigorous evaluations, many others have not at this point.

"Getting to scale" is one of those annoying exhortations of policy makers and foundations that is often invoked at the wrong times and as a put-down rather than a guide for action. At the core of this notion is concern about whether the effects of social innovations can reach larger percentages of the population experiencing such barriers as unemployment and underemployment. Resource constraints require that untapped resources be pursued in scaling up. Consequently, making sense of getting to scale ultimately requires thinking about public and private systems. And systems thinking requires a sense of boundaries, incentives, sustainability, control, and change—not just a leap in magnitude of effort and resources.

Dresser and Rogers articulate a well-developed theory of how to get to scale, which they call *sector-based employment planning*. They argue that organizing firm by firm is limited; rather, groups of firms in sectors should be brought together with their unions and educational and community partners to create clear maps of the labor market for firms and job seekers, including skill standards, career paths, and financing mechanisms. Firms will join together because of labor force needs and because cooperative planning overcomes "free rider" problems associated with firms' taking advantage of the labor force investments of others. Sectoral planning produces not only economies of scale, but also economies of scope, represented in a diversity of programs, and positive externalities in more competitive sectors and regional economies.

Dresser and Rogers extend their argument to show how these sectoral initiatives might influence the public institutions of training. Multiple sectoral employment initiatives in leading regional clusters, these authors believe, can bring coherence to fractured training systems, provide "insider" information about labor demand, leverage private as well as public dollars, and form the core of regional labor market boards.

Several authors in addition to Dresser and Rogers underscore the importance of credentialing as part of scaling up. Credentialing is the establishment of standard performance criteria and skills for particular occupations by employers and educators. Credentials are widely recognized and portable. Nittoli and Giloth discuss the credentialing experience of paraprofessionals during the New Careers movement and the credentialing challenges faced today by less organized community outreach worker occupations. In a different way, completion of the STRIVE program functions as a credential for the "soft skills" demanded by today's employers.

Two other dimensions of systems change that are relevant to scaling up are employer engagement and community colleges. Both have resources and the ability to translate learning from jobs projects into wider arenas of influence. Harrison and Lautsch and Osterman describe STRIVE and Project QUEST, respectively, as performing the human resource function for firms, completing the screening of applicants with due diligence. The organizations' adoption of this role obviously saves money for firms. Dawson argues that jobs projects that invest in frontline workers also save firms money because of reduced turnover—as much as $3,500 per job in home health—and improved service quality. Lautsch and Osterman describe how Project QUEST has engaged employers to design occupations, invent curricula, evaluate and donate equipment, and serve on advisory boards.

Community colleges have resources and know-how that have not been made to connect to low-income communities. Lautsch and Osterman recount the ways in which Project QUEST has used the community colleges in San Antonio to perform training functions, rather than the project's inventing new training mechanisms. The community colleges have had to redesign curricula, lengths of classes, teacher training, case management, basic skills upgrading, and engagement with employers. Community colleges and technical schools are key partners in Dresser and Rogers's version of sectoral employment planning.

Dawson's chapter on Cooperative Home Care Associates provides an example of system change effects within a sector itself. CHCA has become a leading provider of home health services in the New York region, recognized for its high quality, low turnover, and investment in human resources. CHCA also takes some of the credit for recent wage increases among home health firms. Moreover, the organization has become a player in the New York State home health care trade association, lobbying for policies and investments in human resources that will help the entire industry. Here the system is the interrelated public, private, and nonprofit institutions that shape the home health field in one region.

Clavel and Westmont explore getting to scale in the context of politics in the state of Maine. They contrast the "good jobs movement" with the business-as-usual approach to economic development, in which local and state governments give away public incentives to attract companies. Their story shows how a grassroots movement and its partners got on the state agenda, certainly a precondition for getting to scale in the long run.

The good jobs movement includes the JOBS NOW efforts I note in Chapter 1 and the 25 or so livable wage ordinance fights that have taken place around the country in recent years, seeking to establish higher wage floors across the board in cities or states or more narrowly related to firms that contract with public agencies. This organizing includes community organizations, unions, religious congregations, public officials, and even some employers. One of the most effective of these campaigns, summarized by Dewar, has been that carried out by BUILD (Baltimoreans United in Leadership Development), a congregation-based organizing coalition in Baltimore. The livable wage ordinance is one part of BUILD's strategy to organize and support low-wage workers in the service sector. These efforts clearly take on, along with renewal in labor organizing by the AFL-CIO, growing income inequality, wage stagnation, soaring corporate profits, indiscriminate and untargeted public incentives, and now workfare schemes. They try to turn bad jobs into better jobs.

Clavel and Westmont argue that such legislative fights represent only a first move in a longer political process to get voices for good jobs into the policy mix. What they note is striking in their discussion of Maine is the important role of new intellectual, development, and communication capacities to enable a long-term good jobs campaign, one that is highly nuanced and flexible. One example of such a capacity is the Maine Economic Policy Center, modeled after the national Center for Budget and Policy Priorities, which produced studies about the effects of welfare, plant closings, and training policies, and convened several important conferences on high-performance companies and good jobs. Other capacities include women's initiatives in nontraditional employment and the employment and training agreements that Coastal Enterprises, Inc., developed with firms, finding common ground between private investment and good jobs for low-income people. These capacities did not guarantee good jobs advocates victory, but, according to Clavel and Westmont, it got them to the table in a significant way that holds promise for expanding their policy proposals in the future.

Cleveland's Jobs and Work Initiative, described at the outset of this chapter, offers another approach for getting to scale. In this case, leading businesses and philanthropies are at the forefront of the change process without unions; community organizations and the public sector are clearly in a secondary role, at least at this early stage. The civic infrastructure of Cleveland is well organized and focused. How far this effort can go without the full support of other constituencies is an important question. Yet another example of the importance of civic ties for scaling up is found in BUILD's Living Wage campaign.

My opening story about the JOBS NOW conference in Washington, D.C., raises a fundamental caution about the feasibility of getting to scale. There simply are not enough good jobs for everyone seeking them. This presents a crossroads for advocates: whether to make welfare work by getting people attached to bad jobs or to strive consistently for good jobs policies and projects and support attempts to turn bad jobs (incomes) into good jobs. Either decision has cost implications. This also raises the issue of public job creation. The analysis of the New Careers movement and current paraprofessional programs presented in Chapter 9 sheds light on a significant public job creation program. Although my coauthor and I do not think that new federal resources will be available to increase job creation, we do believe that smart negotiations with public agencies and managed-care organizations could produce more and better paraprofessional jobs in the health care field.

These chapters, taken together, provide a reasonably good picture of the current state of job-centered economic development. However, there are omissions. Future work should discuss in more depth renewed labor organizing, municipal approaches to linking workforce and economic development, minority business development, workforce strategies for low-income men and the hard to employ, the overcoming of transportation barriers, and state-level welfare and workforce redesign. The topic of regionalism, although an underlying theme of job-centered economic development, requires more discussion because public policies continue to create divisions between central cities and suburbs that affect the location of firms, access to jobs, and the social isolation of communities. Finally, these chapters suggest that efforts to improve career ladders, to take advantage of networks, and to overcome discrimination are complicated, tenuous, and, in some cases, at a quite early stage of development.

Conclusion

The contributors to this book have sought to explain and document the practice of job-centered economic development, a set of relatively new workforce development, enterprise creation, and policy reforms that have good jobs for low-income job seekers as a central focus. The chapters have situated job-centered economic development in the workings of the contemporary labor market and in relation to other employment strategies, have identified its key characteristics, and have explored several approaches for its replication and bringing it to scale within places, sectors, and the public policy realm. The promise of this strategy is tempered by its small scale, its complex operational demands, the paucity of evaluations of its practice, and the current policy environment.

The environment in which job-centered economic development is occurring offers both opportunities and major challenges. Relatively tight labor markets in most regions make it easier to get the attention of employers and policy makers to support workforce innovations. This is the power of economic incentives to change systems. How long this environment will last is unclear, but most observers suspect it will endure for several years. Yet workforce demographic cycles in industries such as manufacturing and employee turnover costs, more generally among firms, ensure business interest in workforce innovation for years to come. Public institutions, whether schools or traditional economic development agencies, are also more open to workforce innovations given the current economic and policy climate.

This environment is likely to become more complex as employers, government, foundations, and advocacy coalitions increase their efforts to make welfare reform work, critique its failures, establish regional labor force boards, and push for alternatives such as public job creation (Wilson 1996; Center on Budget and Policy Priorities 1997). New national intermediaries are already appearing, such as the Welfare to Work Partnership encouraged by President Clinton, that are proposing to help firms participate in welfare reform. The money of welfare, though reduced in the aggregate, opens up new and attractive options for for-profit companies, economic developers, and many service providers who are positioning themselves.

If national workforce legislation is passed in 1998, it will continue the process of public policy devolution to states. It will consolidate employment training programs into block grants, require employer-driven regional labor force boards, and enable states to develop flexible workforce programs. The Cleveland Jobs and Workforce Initiative is one of many (and one of the best) examples of cities and states getting a head start on establishing a new structure and process of workforce planning and implementation. There are concerns about how this will turn out: The welfare-to-work challenge or the competition for new businesses could absorb the resources of workforce development. A broader challenge is whether states will use this opportunity to integrate economic development, workforce development, and welfare.[2]

Although the focus, flexibility, and local control evident in these changes augurs well for increased investment in job-centered economic development, there are reasons for caution. The atmosphere is one of interest in quick fixes, overnight institutional redesign, caseload reductions, and cost savings. Many of those leading the change process are inexperienced, disconnected from communities, concerned about caseload reduction rather than good jobs, and culpable for the ineffective systems of the past. Already some of the more effective jobs projects are suffering from new funding formulas that do not allow long-term training or that force people into bad jobs. In the short run, these problems are likely to be widespread.

What about the longer term? The real issue, regardless of changes in welfare and workforce laws, is how to enable people to move into good jobs and careers that they can sustain over time. This is the terrain of job-centered economic development. But this will be a long-term struggle, given the trends in the labor market related to wages, career ladders, hiring networks, and discrimination. It is likely that the first and second rounds of welfare changes will achieve modest success at best and more likely painful failures. It is in this context that the lessons and principles of job-centered economic development may achieve a more open hearing, although it is more costly to invest in training for good jobs than in simple job search activities. But this will come to pass only if there is vigilant public scrutiny, cultivation of civic demand, and a good jobs movement calling attention to what can and should be a standard for our society.

The federal role in this policy environment should build upon what the feds are good at. Investing in workforce research and development and promising demonstrations remains important, whether it occurs directly under government auspices or through intermediaries. Likewise, investment in evaluation to determine what really works is essential if we are to distinguish effectiveness from promise. The knowledge gained can then be translated into appropriate benchmarks for workforce development and education of workforce practitioners.

Practitioners of job-centered economic development must consciously transform their collective experience into a field that can participate in the policy world as well as invest in common learning. Many projects are doing this already on an individual basis. A good example is the Sector Employment Development Learning Project, sponsored by the Aspen Institute and six jobs projects, the purpose of which is to engage in a rigorous process of self-evaluation among some of the most experienced practitioners. This practitioner-centered approach to developing a field of practice holds promise for shaping workforce intermediaries, leadership development, and policy agendas.

We began this book with the paradox evident at the Jobs Gap conference in March 1997 of too few good jobs and unfilled good jobs. The contributors to this volume have attempted to make sense of this paradox in their discussions of various aspects of job-centered economic development and the good jobs movement. Despite labor market trends and changing policies, there are reasons for optimism given the level of creativity and innovation found in many effective jobs projects, new forms of experimentation with labor market intermediaries, and community organizing and advocacy around the good jobs issue. It is my hope that the tools, strategies, and examples provided in this volume will contribute to increasing the connections between low-income communities and family-supporting jobs.

Notes

1. For example, foundations are supporting President Clinton's Welfare to Work Partnership as well as efforts by the National Association of Manufacturers and the National Alliance for Business. The Rockefeller Foundation is circulating a white paper on a new intermediary to support a scaling up of welfare-to-work efforts in a limited number of cities. The Ford and Mott Foundations have provided ample capital to Public/Private Ventures in Philadelphia to explore the feasibility of establishing a workforce and community economic development academy. Housing intermediaries such as the Local Initiatives Support Corporation and the Enterprise Foundation are actively exploring their role in the new welfare and workforce development environment. Finally, the Annie E. Casey Foundation's Jobs Initiative is nurturing six local/regional workforce intermediaries. This by no means exhausts the list of intermediaries in action and in the making.

2. A good discussion of national workforce legislation is provided by Baker (1997), who also addresses, in the context of Philadelphia, various local perspectives on the challenge of integrating workforce development, economic development, and welfare policies and investments.

References

Baker, S. 1997. *Restructuring the workforce development system: A wide-angle snapshot.* Philadelphia: 21st Century League.

Berry, D. F. Forthcoming. The Jobs and Workforce Initiative: Building new roads to success in northeast Ohio. *Economic Development Quarterly.*

Center on Budget and Policy Priorities. 1997. *Enhancing income security through public job creation.* Washington, DC: Center on Budget and Policy Priorities.

Cleveland Growth Association. 1997. *Jobs and Workforce Initiative: Building new roads to success.* Cleveland, OH: Cleveland Growth Association.

Connell, J. P., A. C. Kubisch, L. B. Schorr, and C. H. Weiss. 1995. *New approaches to evaluating community initiatives: Concepts, methods, and contexts.* Washington, DC: Aspen Institute.

Harrison, B., and M. Weiss. 1998. *Workforce development networks: Community-based organizations and regional alliances.* Thousand Oaks, CA: Sage.

Melendez, E. 1996. *Working on jobs: The Center for Employment Training.* Boston: Mauricio Gastón Institute, University of Massachusetts.

Okigaki, A. 1997. *Developing a public policy agenda for jobs.* Washington, DC: Center for Community Change.

Replication and Program Services, Inc. 1993. *Building from strength: Replication as a strategy for expanding social programs that work.* Philadelphia: Replication and Program Services, Inc.

Stanfield, R. L. 1997. Just connect. *National Journal,* May 31, 1082-1084.

Wilson, W. J. 1996. *When work disappears: The world of the new urban poor.* New York: Knopf.

Index

Access, Support and Advancement Partnership (ASAP), 147, 150
Accountability:
   corporate, 197, 199, 201, 203
   program, 179, 180
AFSCME (American Federation of State, County, and Municipal Employees), 183, 184, 185
Alinsky, Saul, 183, 217
American Federation of State, County, and Municipal Employees (AFSCME), 183, 184, 185
Anderson, Elijah, 34
Apprenticeship programs, 74
ASAP (Access, Support and Advancement Partnership), 147, 150
Aspen Institute, 247
Assessment of participants, 105, 113-117. *See also* Evaluation
Assets approach, 187-191
AT&T, 33
Attitudinal training, 140-141, 148-149. *See also* Soft skills

Baltimore, MD:
   BUILD in, 180, 183-186, 243
   Healthy Start in, 160, 162

Baltimoreans United in Leadership Development (BUILD), 180, 183-186, 243
Banks, 25
Barlett, D. L., 199
Becker, Gary, 32
Benefits, employment, 22-24, 25, 125, 126
Bergmann, Barbara, 21
Bishop, John, 30-31
Blackwood, D. L., 46
Blom, Eric, 198, 199
Blum, Harrison Michael, 146
Boisvert, Rene, 204
Boston, MA:
   CHCA in, 123, 128-129, 136, 240
   STRIVE in, 137, 145
Branch plant manufacturing, 196, 197
Bridges to Work project, 56, 57, 58
Brigham, Alan, 201, 203
BUILD (Baltimoreans United in Leadership Development), 180, 183-186, 243
Business cluster analysis, 87, 95-98, 99-101
Business incentive programs, 43-50, 198, 206-207
Business incubators, 190
Business success, 132-135
Bus-Rail Link program, 56
Buyouts, employee, 204-205

249

California, 8, 48-49, 215
Camden neighborhood, 187
Canales, Jim, 142
Cappelli, P., 28, 29
Career ladders, 7, 26, 28
    clarity needed in, 65, 70-71
    paraprofessionals on, 155, 160, 164, 166, 168, 170
    sectoral strategies for, 72, 76-78
    STRIVE and, 138, 146-147
    workplace changes and, 65-71
Carl D. Perkins Vocational and Applied Technology Act of 1990, 33
Carlson, Virginia, 110-111
Carmona, Robert, 139, 140, 141, 142, 143, 145, 146
Cavanaugh, Mike, 204, 205
CDBGs (Community Development Block Grants), 44
CEI (Coastal Enterprises, Inc.), 206-208, 209, 244
Census data, 91, 109-110, 111
Census of Manufactures, 91
Census of Service Industries, 91
Center for Employment Training (CET), 8-9, 31, 53, 215
    effectiveness of, 234, 238, 239, 240
Center on Wisconsin Strategy (COWS), 72, 76, 77
CET. *See* Center for Employment Training
CETA (Comprehensive Employment and Training Act), 51, 155, 219
CHAs (community health advisers), 158-162
CHCA. *See* Cooperative Home Care Associates
CHCB. *See* Cooperative Home Care of Boston
Chicago, IL:
    Gautreaux Assisted Housing Program in, 55-56, 58
    Grand Boulevard program in, 152, 153
    Job Oasis in, 57
    labor force data on, 110-111
    Project Match in, 113, 116-117, 118
    STRIVE in, 137, 145, 146, 240
    Suburban Job Link Corporation in, 56-57, 58
Child-care programs, 162, 163, 164-168
Childhood development programs, 164-168, 170
Children in poverty, 4

Citizen involvement, 179-193
Clark Foundation, 143-144
Cleveland, OH, 234, 244, 246
Cleveland-Akron, OH, 107-108
Cluster analysis, 87, 95-98, 99-101
Coastal Enterprises, Inc. (CEI), 206-208, 209, 244
College education, 20, 23, 25. *See also* Community colleges; Educational level
Collins, Susan, 199
Communications Workers of America, 33
Communities:
    development incentives for, 48-50
    institutional change in:, 229-232
    labor-based design and, 133-134
    re-membering of, 191-193
    strong ties in, 149, 152, 179-193, 229-230, 239
Communities Organized for Public Service (COPS), 217, 229, 230, 235, 239
Community analysis, 180-182, 192-193
    by BUILD, 186
    by Interfaith Action, 187-190
Community Career Ladder Project, 72, 76-78
Community colleges:
    emerging role of, 34
    institutional change and, 225-229
    Project QUEST and, 218, 223, 225-229, 231, 242-243
Community Development Block Grants (CDBGs), 44
Community first! ethic, 182
Community health advisers (CHAs), 158-162
Community involvement options, 179-193, 208
Community Reinvestment Act, 167, 172
Community Talent Inventory, 180, 188-190, 191
Company-based training, 31-33, 132. *See also* Training
Comprehensive Employment and Training Act (CETA), 51, 155, 219
Cooperative Extension Service, 159
Cooperative Health Care Network, 123-136
Cooperative Home Care Associates (CHCA), 123, 238, 240, 243
    changes in, 129, 136
    labor market profiling and, 113, 114-116, 118
    SEEDCO and, 162

start-up of, 126-127
Cooperative Home Care of Boston (CHCB), 123, 128-129, 136
Cooperatives, worker-owned, 123-136
COPS (Communities Organized for Public Service), 217, 229, 230, 235, 239
Core competencies, 130
Costs:
   development incentives and, 43-50
   flexible work organization and, 26-29
   of dead-end jobs, 68
   of pay and benefits, 23
   of placement, 145
   of retention, 145
   of turnover, 135
   of workforce investment, 135
   per participant, 222
Counseling, in Project QUEST, 218-219, 223
County Business Patterns, 89, 102(n3)
COWS (Center on Wisconsin Strategy), 72, 76, 77
Creaming, 224
Credentialing, 169-170, 242
Crowding hypothesis, 21
Cruickshank, Joseph, 142
Culture, organizational, 130

Dabney, D. Y., 48
Data collection:
   for labor market profiles, 105-118
   for regional economic analysis, 88-98
Dead-end jobs, 67-68
Demand-side strategies, 4-5
   HPWOs and, 203
   regional economic analysis and, 85-87
Demographic characteristics, 91, 109-110, 111
Design elements, 132-135
Design stage of projects, 106
Detroit, MI:
   Focus Hope in, 215
   PEI in, 108-109, 113
Development incentives, 43-50, 198, 206-207
Diesel mechanics program, 226-228
Differently abled persons, 208
Disabled persons, 207-208
Discrimination, 7, 30, 140-141
Dispersal, residential, 54-56
Double social utility, 153
Dowall, D. E., 48-49

Doyle, Frank, 1
Dual labor market theory, 5
Dual model design, 132

Early childhood development, 164-168, 170
Earned Income Tax Credit (EITC), 58
Earnings. *See* Income
East Harlem, NY, 137
Economic analysis, regional. *See* Regional economic analysis
Economic development:
   goal of, xi-xii
   jobs linked with, xi-xii
   multiple dimensions of, 2
   See also Job-centered economic development
Economic Opportunity Act of 1964, 155
Education:
   paraprofessionals in field of, 156, 162-168
   reliability vs. quantity of, 21-23
Educational level:
   as program entrance requirement, 113
   gender and, 4, 21
   income and, 4, 21, 23
   labor market profiles of, 107
   low-skilled labor and, 20
   program evaluation and, 220, 222-223
   race and, 4, 21
   training and, 30-31, 53
Educational systems, impact on, 225-229, 231-232
Effectiveness, lessons learned about, 237-239. *See also* Evaluation
EITC (Earned Income Tax Credit), 58
Eligibility criteria, 147, 150, 217. *See also* Selection of participants
Employee buyouts, 204-205
Employee-owned cooperatives, 123-136
Employer-centered training, 31-33, 132. *See also* Training
Employers:
   development incentives for, 43-50, 198, 206-207
   in Project QUEST, 218, 223, 225, 230-231
   institutional change and, 225, 230-231
Employer Wage Credit program, 50
Employment and training agreements (ETAGs), 206-208, 209
Employment cost index, 23

Employment system, changes in, 25-29, 65-71
Empowerment of participants, 141
Empowerment zones, 44, 48, 49-50
Enterprise zones, 44, 48-49
Entrance requirements. *See* Eligibility criteria; Selection of participants
ETAGs (employment and training agreements), 206-208, 209
Evaluation:
    case studies and sources for, 113-117
    of paraprofessional programs, 160, 165-166
    of Project QUEST, 215-216, 220-232, 238
    of STRIVE, 147-148
    See also Effectiveness; Failures; Information gathering
Evaluation literature, limitations of, 45, 215
Evans, Seth, 128
Experimental evaluation methods, 214, 215

Failures, in conventional approaches, 42-59, 191-192, 237-238
FAME (Finance Authority of Maine), 197, 198, 210
Family support services, 156, 168, 170
Ferguson, Ronald, 20
Finance Authority of Maine (FAME), 197, 198, 210
Financial aid, in Project QUEST, 218, 219
Financial industry, 25, 77
Fishgold, G., 46
Flexible work organization, 26-29
Focus Hope, 215
Follow-up systems, 8, 137, 142-145, 149-150
Ford Foundation, 216
Fragmentation, program, 7-8, 169-170
Frey, Michael, 143

Garber, Andrew, 198, 199
Gautreaux Assisted Housing Program, 55-56, 58
Gender:
    education and, 4, 21
    government jobs and, 24
    home health care and, 125, 131, 133
    income and, 4, 23, 52
    informal networks and, 69
    Maine programs and, 205, 209

    STRIVE programs and, 145
General Motors, 33
Geographic targeting, 48-50
Geography:
    barriers and, 6
    of economic analysis, 88, 93-94, 96
    of group membership, 191
Gertz, Lyle, 138, 139
Gildart, Kevin, 197
Glass ceilings, 69
Good jobs movement, 196-210
Gordon, Scott, 127
Government jobs, 24, 25, 66
Grand Boulevard program, 152, 153
Grassroots involvement, 179-193, 208

Hard skills, 6
Harrison, Lorenzo, 142, 145
Hartwell, Samuel L., 138, 139, 142
Hastedt, Chris, 199, 200
Hathaway, John, 199
HCA (Home Care Associates), 123, 127-128, 136
Head Start, 163, 164, 165, 168
Health care industry, 25, 52, 77
    institutional change and, 225
    paraprofessionals in, 123-136, 156-162
Healthy Start, 159, 160, 161, 162
HHA (Home Health Aide) Demonstration, 52
High-performance workplace organizations (HPWOs), 202-204, 206, 208, 209
High school/college partnerships, 228-229
Hillard, Michael, 198, 203, 208, 210
Hiring networks, 6-7, 69-70
Holzer, Harry, 21
Home Care Associates (HCA), 123, 127-128, 136
Home Health Aide (HHA) Demonstration, 52
Home health care, 52, 123-136, 158
Hornbeck, J. F., 49
Horton, Frank, 139, 140
Housing and Urban Development (HUD), U.S. Department of, 55, 56, 57
Housing-based strategies, 9, 54-56
HPWOs (high-performance workplace organizations), 202-204, 206, 208, 209
HUD (Housing and Urban Development), U.S. Department of, 55, 56, 57

Hughes, Mark Allen, 57, 58
Human service paraprofessionals, 123-136, 154-172

IAF (Industrial Areas Foundation), 183, 184, 217, 229-230, 239
Impact. *See* Effectiveness; Evaluation; Institutional change
Implementation stage of projects, 106, 113-117
Incentives, development, 43-50, 198, 206-207
Income:
　declines in, 4, 22-24
　failed strategies and, 47, 52, 53, 56, 57-58
　financial aid as, 218, 219
　for paraprofessionals, 159, 160, 165, 166-167, 168, 171
　in BUILD program, 184
　in Cooperative Health Care Network, 125, 126
　in Project QUEST, 218, 219, 221
　in regional economic analyses, 89
　labor market restructuring and, 21, 22-24, 25, 27, 29
　livable wage as, 12(n1)
　Maine politics and, 199, 201, 204
　sectoral strategies and, 75
Incremental Ladder to Economic Independence, 117
Incubators, business, 190
Indian Health Service, 159, 170
Individuals With Disabilities Act, 163, 166, 172
Industrial Areas Foundation (IAF), 183, 184, 217, 229-230, 239
Industrial Development Agency, 46
Industrial development strategies, inadequacies of, 44-50
Industrial Outlook, 93
Industrial revenue bonds, 44
Informal hiring networks, 6-7, 69-70
Information gathering:
　for labor market profiles, 105-118
　for regional economic analysis, 88-98
　See also Evaluation
Institutional change, 224-230
Insurance industry, 77, 78
Interfaith Action, 180, 187-191
International comparisons, 31, 51, 74, 232

Jackson, Lawrence, 139
Job-centered economic development:
　characteristics of, 9, 10, 236, 246-247
　components of, 2-3
Job ladders. *See* Career ladders
Job Oasis, 57
Job Opportunities and Basic Skills, 33
Job Opportunities in the Business Sector (JOBS), 53
Job-Ride program, 56
Jobs:
　collecting data on, 106-109
　economic development linked with, xi-xii
　multiple dimensions of, 2
　organizing for, 179-182, 183-186
Jobs and Workforce Initiative (JWFI), 234, 244, 246
Job security, changes in, 26-29
Jobs for Youth, 147
Jobs Gap conference, 1-2, 247
JOBS (Job Opportunities in the Business Sector), 53
JOBS NOW Coalition, 1, 243
Job Training Partnership Act (JTPA), 8, 33, 51, 52
　labor force profiling and, 112
　stipends and, 219
　STRIVE and, 142
JTPA. *See* Job Training Partnership Act
Juvenile justice system, 156, 168
JWFI (Jobs and Workforce Initiative), 234, 244, 246

King, Angus, 197, 198, 201
Kretzmann, J., 188

Labor-based design, 133-135
Labor buyouts, 204-205
Labor force:
　critical indicators of, 109, 111, 113
　data collection on, 105, 109-112
　defining targets within, 112-113
　strategies aimed at, 50-58
Labor-management collaboration, 73-74
Labor markets:
　administration problems in, 79-82
　changes in, 19-36, 65-71, 225, 230-231, 237

clarity needed in, 65-71, 237
day-to-day impact on, 231, 232
devolution of, 25-29
lessons learned about, 237
outcomes and barriers in, 3-8
profiling of, 105-118
Labor unions, 7, 33, 66
  in BUILD programs, 183, 185
  in sectoral strategies, 73, 74
  Maine politics and, 205, 208
  paraprofessionals and, 167-168
Ladders:
  career. *See* Career ladders
  of participant steps, 117
Lay advisers, 152. *See also* Paraprofessionals
Layoffs, 27-28
Leased workers, 27
Liberty, Michael, 204
Livable wage, 12(n1)
Living wage campaign, of BUILD, 184-186
Local Initiatives Support Corporation, 9
Location quotient analysis, 93, 103(n6)
Louisville, KY, 235
Low-skilled labor, 20-21, 58
Lynch, R. G., 46

Macpherson, C. B., 195
Madison, WI, 76-78
Maine, 196-210, 239, 243
Maine Center for Economic Policy (MCEP), 198, 199, 209-210, 243
Maine Displaced Homemakers (MDH), 205
Managed-care companies, 161-162
Managers, in replicated programs, 130-131
Mangum, Garth, 36
Manufacturing:
  Maine politics and, 196, 197-210
  sectoral strategies in, 72-77
Manufacturing Extension Program (MEP), 74
Massachusetts, 235. *See also* Boston
Match, Project, 113, 116-117, 118
MCEP (Maine Center for Economic Policy), 198, 199, 209-210, 243
McKnight, J., 188
MDH (Maine Displaced Homemakers), 205
Medicaid, 129, 161, 171-172
Medicare, 125, 126
Men. *See* Gender
Mental health services, 156, 168

MEP (Manufacturing Extension Program), 74
Mergers, 25
Meter, Ken, 187, 188
Metro Alliance, 217, 229, 230, 235, 239
Metropolitan statistical areas (MSAs):
  labor market profiles and, 110
  primary (PMSAs), 102(n1)
  regional economic analyses and, 88, 91, 93
MFSP (Minority Female Single Parent) Demonstration, 52, 53
Milwaukee Jobs Initiative (MJI), 72, 74-76
Mincy, Ronald, 146, 147
Minneapolis, MN, 180, 187
Minority Female Single Parent (MFSP) Demonstration, 52, 53
Miranda, Sal, 190
Mission, organizational, 130, 135, 143
MJI (Milwaukee Jobs Initiative), 72, 74-76
Mobility strategies, 54, 56-58
Monitoring, program, 113-117
More-and-better-programs response, 192
More-public-private-partnerships response, 192
Morris, Aldon, 181
Moving to Opportunity (MTO) program, 55, 56
MSAs. *See* Metropolitan statistical areas
MTO (Moving to Opportunity) program, 55, 56

Neighborhood Balance Sheet, 180-181, 187
Neighborhoods:
  incentives for, 48-50
  strategies for, 133-134
  See also Communities; Community analysis
Networks:
  community, 181-193
  for hiring, 6-7, 69-70
  in sectoral partnerships, 72
  of Cooperative Health Care, 123-136
  of STRIVE programs, 143-144
  training embedded in, 9
New Careers movement, 153, 154-156, 171, 238, 242, 244
New York (City):
  CHCA in, 113, 114, 123
  STRIVE in, 137, 142, 143, 147
New York (State), 46

Nussbaum, Karen, 2

Open entry/open exit instruction, 226
Oregon, 235
Organizational characteristics, 130
Organizing for jobs, 179-182, 183-186
Outreach workers, 152. *See also* Paraprofessionals
Outside-in response, 191
Outsourcing, 27, 66

Paraprofessional Healthcare Institute, 124, 127, 131
Paraprofessionals:
　barriers to, 160-161, 166-168
　definition of, 156
　effectiveness of, 160, 165-166, 244
　in Cooperative Health Care Network, 124-136
　in New Careers movement, 153, 154-156, 238
　job creation for, 156-168
　policy issues for, 168-171
　professionals and, 169-170
　rationales for, 152-154
　roles of, 157-158, 169
Partnership for Economic Independence (PEI), 108-109, 113
Partnerships:
　high school/college, 228-229
　limitations of, 192
　public-private, 192, 195, 196-210
　sectoral, 71-82
Part-time schedules, 27
Pay. *See* Income
Payback period, 222
Pearl, Arthur, 154
PEI (Partnership for Economic Independence), 108-109, 113
Permanent membership groups, 191
Personal Responsibility and Work Opportunity Reconciliation Act of 1996, 2, 4, 12(n3)
Philadelphia, PA, 123, 127-128, 136, 240
Phillips, Ron, 206
Pinewood Estates, 107-108
Pingree, Rochelle, 197, 199, 200
Pittsburgh, PA, 137, 142, 145
Planning:
　data gathering for, 105-113
　regional economic analysis for, 85-102
PMSAs (primary metropolitan statistical areas), 102(n1)
Postal zip codes, 91, 93, 107
Postplacement support, 8, 137, 142-145, 149-150
Poverty rates, 4
Powell, Peggy, 126, 127
Primary metropolitan statistical areas (PMSAs), 102(n1)
Primary research:
　in labor market profiles, 108-109, 111-112, 118
　in Project QUEST, 216, 220, 223
　in regional economic analyses, 91-92, 94, 97-98
　in talent inventory, 188-190
Professionals, and paraprofessionals, 169-170
Profiling:
　of labor markets, 105-118
　of regional economy, 88-91
Profitability, in dual model design, 132
Project Match, 113, 116-117, 118
Project QUEST, 113, 118, 235, 239, 242-243
　data gathering for, 111-112
　evaluation of, 215-216, 220-232, 238
　institutional change and, 224-230
　program design of, 217-220
Promotion opportunities. *See* Career ladders
Public Employment Service Act of 1970, 155
Public sector:
　partnerships with, 192, 195, 196-210
　sectoral strategies and, 78-82
　See also Government jobs

Quality jobs movement. *See* Good jobs movement
Quasi-experimental methods, 214
QUEST. *See* Project QUEST

Race:
　discrimination and, 7, 30
　education and, 4, 21
　government jobs and, 24
　income and, 4
　informal networks and, 69
　skills and, 20, 30
　unemployment rates and, 4

Racism, 140-141
Rainwater, Lee, 34
Random-assignment methods, 214, 216, 224
Ransom, Avis, 184, 185
Recession, economic, 19-20
Regional advantage, 93
Regional economic analysis:
    cluster analysis in, 87, 95-98, 99-101
    community analysis and, 182
    firm identification in, 91-92
    profile analysis in, 88-91
    sector analysis in, 87, 92-95
    usefulness of, 85-87
Re-membering of community, 191-193
Replication:
    design elements of, 132-135
    lessons learned about, 129-131, 240-241
    of CHCA model, 123, 127-129, 132-135, 240
    of STRIVE model, 141-146, 240
    principles of, 241
Residential dispersal, 54-56
Retention, 143, 144-145, 148, 217
Return on investment, 222
Reverse commuting, 56-58
Riessman, Frank, 154, 155
Rodman, Tom, 139
Rosenbaum, J. E., 55-56

Salary. *See* Income
San Antonio, TX, 111-112, 215-232, 235
San Jose, CA, 8
Scaling up of projects, 72, 241-245
Schooling, years of, 22, 23. *See also* Educational level
Schools, paraprofessionals in, 162-168
School to Work Opportunities Act, 33
School-to-work programs, 74
Screening. *See* Eligibility criteria; Selection of participants
Seattle Jobs Initiative, xii-xiii
Secondary data:
    in labor market profiles, 107-108, 109-110, 111
    in Project QUEST, 216, 220, 223
    in regional economic analyses, 88-91, 93, 97
Sector advantage, 93
Sectoral employment strategy, 133

Sectoral strategies:
    effectiveness of, 238-239
    evolution toward, 215
    examples of, 72-78
    labor market administration and, 79-82
    scaling up as, 72, 241-242
    theory of, 71-72
Sector analysis, 87, 92-95
Sector Employment Development Learning Project, 247
SEEDCO, 162
Seguino, Stephanie, 209
Selection of participants, 113, 217-218, 222-224. *See also* Eligibility criteria
Self-interest, of local people, 179-193
Self-reliance, community, 179-193
Self-sufficiency, 147, 150, 152
Self-sufficiency ladder, 117, 118
Seniority systems, 26, 28, 66
Service sector, career ladders in, 66, 67
Shift-share analysis, 93, 104(n7)
SIC (Standard Industrial Classification), 89, 91, 93, 102(n2)
Silicon Valley, 235
Skills:
    basic, 20, 22
    employer priorities on, 21-23
    hard, 6
    inventory of, 180, 188-190, 191
    labor market profiles of, 107
    mid-level jobs and, 68-69
    mismatches in, 5
    mobility programs and, 57, 58
    myth about, 20
    sectoral strategies and, 73, 77
    sector analysis and, 95
    soft, 5, 6, 30, 137, 140-141, 148-149
    work reorganization and, 65, 66
    See also Training
Snowe, Olympia, 199
Social networks, 181, 193
Soft skills, 5, 6, 30
    STRIVE and, 137, 140-141, 148-149
South Bronx, NY, 113, 114, 123
South Eastern Pennsylvania Transportation Authority, 56
Spanish Coalition for Jobs, 111
Spatial mismatch, 6
Stages of employment projects, 106

Standard Industrial Classification (SIC), 89, 91, 93, 102(n2)
Standardized tests, 113, 114
Standards of practice, 169-170
Stanley M. Isaacs Neighborhood Center, 147
Start-ups:
  of CHCA, 126-127
  of STRIVE, 138-141
  See also Replication
Steele, J. B., 199
St. John, Kit, 201
St. Phillips College, 226, 227
STRIVE, 137-150, 238, 239, 240, 242
Suburban employment, links to, 54-58
Suburban Job Link Corporation, 56-57, 58
Supervision, of paraprofessionals, 170-171
Supin, Rick, 126, 127
Supply-side strategies, 85, 203
Supported Work Demonstration (SWD), 52-53
Survival skills, 139. *See also* Soft skills
SWD (Supported Work Demonstration), 52-53

Talent inventory, 180, 188-190, 191
Targeted development incentives, 46-50
Targeted Jobs Tax Credit (TJTC), 46-47, 50
Targeting, effectiveness of, 238
Target population:
  data collection on, 109-112
  defining of, 112-113
Tax incentive programs, 43-44, 46-50, 198. *See also* Earned Income Tax Credit
Technology:
  New Careers movement and, 154
  sectoral strategies and, 73, 74, 77
Temporary service agencies, 7, 27, 67, 186
Temporary-to-permanent placement strategy, 128
Terminations, 217. *See also* Retention
Tershy, Russ, 9
Theodore, Nikolas, 110-111
Tilly, Chris, 30
TJTC (Targeted Jobs Tax Credit), 46-47, 50
Training:
  company-based, 31-33, 132
  dead-end jobs and, 68
  effectiveness of, 30-31, 34-36
  evaluation tools for, 214-216
  follow-up system and, 142-145, 149-150
  fragmented, 7-8, 169-170
  institutional change and, 224-230
  investment in, 31-32, 68-69
  labor market profiles of, 107
  Maine politics and, 201, 205-208
  mid-level jobs and, 68-69
  participant assessment in, 105, 113-117
  strategy for, attitudinal, 140-141, 148-149
  strategy for, conventional, 50-54
  strategy for, new approaches, 34-36
  strategy for, sectoral, 73-78, 80-82
  welfare reform and, 52, 53
  See also Skills
Transportation-based strategies, 54, 56-58
Trickle-down theory, xii, 44, 184
Turnover, employee, 125, 135, 167

UDAGs (Urban Development Action Grants), 44, 45-46
Unemployment, 3-4, 51
Unions. *See* Labor unions
United Auto Workers, 33
Urban Development Action Grants (UDAGs), 44, 45-46

Vail, David, 198, 203, 206, 208, 210
Validity, in evaluation, 222-224
Value-adding enterprises, 90-91, 238
Vocational and Applied Technology Act of 1990, 33
Vocational training partnerships, 228-229
Volunteers, 159

Wages. *See* Income
Walton, Alice, 204
War on Poverty, 153, 154
Welfare reform:
  Maine politics and, 197, 200, 207
  opportunities and, 245-246
  paraprofessionals and, 153
  training and, 52, 53
Welfare to Work Partnership, 245
White, Richard, 227
Williamson, B., 198
WIN (Work Incentive) program, 53
Wisconsin, 56, 72-78
Wisconsin Regional Training Partnership (WRTP), 72, 73-75, 76

Women:
  glass ceilings for, 69
  in home health care industry, 125, 131, 133
  in Maine programs, 205, 209
  in MFSP program, 52, 53
  in STRIVE programs, 148
  See also Gender
Wong, J. L., 156
Worker buyouts, 204-205
Worker-owned cooperatives, 123-136
Work ethic, 5, 137. *See also* Soft skills
Workfare organizing, 185. *See also* Organizing for jobs
Workforce Boards, 235
Workforce development, 29, 238. *See also* Labor force; Training

Work Incentive (WIN) program, 53
Work organizations:
  changes in, 25-29, 65-71
  clarity needed in, 65, 70-71, 237
  heterogeneity in, 27
  high-performance, 202-204, 206, 208, 209
WRTP (Wisconsin Regional Training Partnership), 72, 73-75, 76

Youngblood, Johnny Ray, 191

Zemsky, Robert, 22
Zip codes, 91, 93, 107

ABOUT THE AUTHORS

Brian Bosworth is a principal with Regional Technology Strategies, Inc. He is a graduate of Dartmouth College and the Fletcher School of Law and Diplomacy at Tufts University, and he has more than 30 years' experience in economic and business development. He served as President of the Indiana Development Council, Inc., from 1984 to 1988, after serving 8 years in Indiana state government.

Peggy Clark is Director of the Economic Opportunities Program at the Aspen Institute, a program whose two major projects are the Self-Employment Learning Project and the Sectoral Employment Development Learning Project. Previously, she worked on small business and microenterprise programs for the Ford Foundation and the Save the Children Federation. She is the author of numerous publications and holds a master's degree in international economics and Latin American studies from the Johns Hopkins School of Advanced International Studies.

Pierre Clavel, Ph. D., teaches in the Department of City and Regional Planning at Cornell University. He is the author of *Progressive Cities*; coauthor, with Norman Krumholz, of *Reinventing Cities: Equity Planners Tell Their Stories*; and coeditor of *Harold Washington and the Neighborhoods: Progressive City Government in Chicago, 1983-1987*. He is currently researching the progressive role of community development corporations in three places.

Steven L. Dawson is President of the Paraprofessional Healthcare Institute, a nonprofit training and replication affiliate of Cooperative Home Care Associates in South Bronx, New York. Previously, he served as founding Executive Director of the Industrial Cooperative Association in Boston and as Field Director for the National Association for the Southern Poor, a rural community organizing initiative in Petersburg, Virginia.

Thomas R. Dewar, Ph.D., is Senior Project Associate at Rainbow Research, Inc., of Minneapolis. He has taught and done action research on the role of mutual aid and informal networks in community building, the social and economic organization of neighborhoods, and program evaluation. Trained as a sociologist, he was on the faculty of the Hubert H. Humphrey Institute of Public Affairs, University of Minnesota, for 15 years.

Laura Dresser, Ph.D., is Research Director at the Center on Wisconsin Strategy, a policy research institute at the University of Wisconsin–Madison. She is trained as a labor economist, and her principal research interests are in occupational sex segregation, human capital theory, and service sector productivity. She is the lead author of the recent *State of Working Wisconsin,* a comprehensive analysis of wage and income trends in the Wisconsin regional economy.

Robert P. Giloth, Ph.D., is a Senior Associate with the Annie E. Casey Foundation in Baltimore, where he manages the six-city Jobs Initiative. He has directed community development corporations in Chicago and Baltimore and served as Deputy Commissioner for Economic Development under Mayor Harold Washington in Chicago.

Bennett Harrison, Ph.D., is Professor at the Robert J. Milano Graduate School of Management and Urban Policy of the New School for Social Research in New York. He is the author of a number of journal articles and books, including *Lean and Mean: The Changing Landscape of Corporate Power in the Age of Flexibility* and *Workforce Development Networks: Community-Based Organizations and Regional Alliances,* with Marcus Weiss.

Lorenzo D. Harrison is Deputy Executive Director of East Harlem Employment Service/STRIVE, Inc., in East Harlem, New York. He started with STRIVE in 1988 as a job developer. His current responsibilities include administration, management, fund development, program development, public relations, and technical assistance. He holds a master's degree in public administration from New York University.

Amy J. Kays is Associate Director of the Economic Opportunities Program at the Aspen Institute and has worked on the institute's self-employment and sectoral employment projects. Previously, the author of numerous publications, she conducted research on the economic and employment impacts of manufacturing globalization. She holds an M.S. in community and regional planning from the University of Texas at Austin.

Brenda A. Lautsch is a doctoral candidate in the Industrial Relations and Human Resource Department of the MIT Sloan School of Management. She is currently engaged in research on the growth of contingent work and its implications for firms and workers. She has also conducted research on inequality and work, impacts of team-based work systems, and the use of alternative dispute resolution to resolve civil rights disputes. Her experience has also included a faculty appointment at the University of Regina.

Elizabeth J. Mueller, Ph.D., is Assistant Professor and Senior Research Associate in the Community Development Research Center, Robert J. Milano Graduate School of Management and Urban Policy at the New School for Social Research. She is coauthor of *From Neighborhood to Community: Evidence on the Social Effects of Community Development.* Her current research centers on strategies for increasing stable employment in low-income communities and the regional differences in community development activity. She is a member of the evaluation team of the Annie E. Casey Foundation's Jobs Initiative.

Janice M. Nittoli is a Senior Associate at the Annie E. Casey Foundation, where she leads the foundation's efforts to reform frontline social service delivery. Previously, she led the National Center for Health Education, after working for 8 years in various human service positions in New York City government, including Assistant Commissioner in the Department of Health. She holds a master's degree in public and international affairs from the Woodrow Wilson School of Princeton University.

Paul Osterman, Ph.D., is Professor of Human Resources and Management at the Sloan School, Massachusetts Institute of Technology. He is the author of *Getting Started: The Youth Labor Market* and *Employment Futures: Reorganization, Dislocation, and Public Policy;* coauthor of *The Mutual Gains Enter-*

prise: Forging a Winning Partnership Among Labor, Management, and Government; and editor of *Internal Labor Markets* and *Broken Ladders: Managerial Careers in the New Economy*. He has been a Senior Administrator of Job Training Programs for the Commonwealth of Massachusetts and has consulted widely to government agencies, foundations, community groups, and public interest organizations.

Joel Rogers, Ph.D., is Professor of Law, Political Science, and Sociology at the University of Wisconsin–Madison and Director of the Center on Wisconsin Strategy. He has written widely on democratic theory, labor relations, and American politics and public policy. His most recent books are *Works Councils: Consultation, Representation, and Cooperation in Industrial Relations, Associations and Democracy,* and *A Strategy for Labor.* He serves as a consultant to the Democratic minority at the Labor and Human Resources Committee of the U.S. Senate, is a cofounder of Sustainable America, and is chair of the New Party.

Alex Schwartz, Ph.D., is Assistant Professor and Senior Research Associate in the Community Development Research Center, Robert J. Milano Graduate School of Management and Urban Policy at the New School for Social Research. He is coauthor of several publications on the management of low-income housing owned by nonprofit organizations. His most recent research focuses on the implementation of Community Reinvestment Act agreements. He is also part of the evaluation team of the Annie E. Casey Foundation's Jobs Initiative.

Marcus Weiss, Esq., is President of the Economic Development Assistance Consortium, a Boston-based consulting firm that works with community-based organizations, foundations, government agencies, and financial institutions across the nation. He serves as counsel to the Massachusetts Association of CDCs and as a regular consultant to the National League of Cities, the U.S. Department of Housing and Urban Development, and the U.S. Department of Commerce. He is coauthor, with Bennett Harrison, of *Workforce Development Networks: Community-Based Organizations and Regional Alliances.*

Karen Westmont is in the Master's Program in City and Regional Planning at Cornell University. Previously, she worked for more than a decade in California on housing, community advocacy, and grassroots electoral campaigns.